Reference Serie

Computer Security Sourcebook

Basic Information for General Readers about Computer, Internet, and E-mail Security, Including Information about Data Backups, Firewalls, Passwords, Virus Protection, Sensitive Data Encryption, Internet Filtering, E-mail Monitoring and Security, Children's Online Privacy and Security, Privacy Rights and Policies, Online Monitoring, and More

Along with a Glossary of Related Terms and Resources for Further Information

Edited by Wilma R. Caldwell. 514 pages. Index. 2003. 0-7808-0648-4. $68.

A reported 52 percent of U.S. computer users are connected to the Internet, and the *Computer Industry Almanac* expects to see more than 765 million Internet users worldwide by the end of the year 2005. Security concerns associated with the Internet and other advanced technologies are rising. Information on these concerns is often widely scattered.

The *Computer Security Sourcebook* brings together a wide range of facts, figures, and resources to better assist the general reader in navigating today's computer security issues. The book covers the basics of computer, Internet, and e-mail security to help consumers protect themselves from fraud, privacy invasion, hackers, lost data, computer viruses, and other security issues associated with computer use.

Travel Security Sourcebook

Basic Information for General Readers about Security Issues Related to Travel, Including Information about Safety in Airports and on Airplanes, Preparing for and Responding to Crises and Emergencies while Traveling Abroad, Avoiding Health Risks, and Protecting Money and Possessions while Traveling

Along with Tips for Safe Traveling Related to Children, Students, Women, Businesspeople, and the Elderly, Statistics, a Glossary of Related Terms, and Resources for Further Information

Edited by Chad T. Kimball. 600 pages. Index. 2003. 0-7808-0617-4. $68.

The *Travel Security Sourcebook* provides the latest information on safeguarding the security of the traveler, including airport and airline safety information, responding to crises and emergencies abroad, preserving one's health during travel in foreign countries, protecting money and possessions while traveling, and other travel security concerns.

Information in this *Sourcebook* will be useful for every type of traveler, including international, domestic, business, and leisure travelers. It will help them prepare for unexpected emergencies and crises as well as the normal occurrences involved with travel, whether by airplane, train, subway, bus, or automobile. Tips on various travel matters as well as sources of assistance are included throughout the book.

Communications Security SOURCEBOOK

Security Reference Series

Communications Security SOURCEBOOK

Basic Information for General Readers about Cell Phone and Wireless Communication Security, Telephone Company Security Issues, Telephone Slamming and Cramming, Long Distance Telephone Scams, Telemarketing Fraud and Other Nuisances, Wiretapping and Eavesdropping, and More

Along with Glossaries of Related Terms and Resources for Further Information

Edited by Wilma R. Caldwell

615 Griswold Street • Detroit, MI 48226

Bibliographic Note

Because this page cannot legibly accommodate all the copyright notices, the Bibliographic Note portion of the Preface constitutes an extension of the copyright notice.

Edited by Wilma R. Caldwell

Security Reference Series

Chad T. Kimball, *Series Editor*
Peter D. Dresser, *Managing Editor*
Elizabeth Barbour, *Permissions Associate*
Dawn Matthews, *Verification Assistant*
Laura Pleva Nielsen, *Index Editor*
EdIndex, Services for Publishers, *Indexers*

* * *

Omnigraphics, Inc.

Matthew P. Barbour, *Senior Vice President*
Kay Gill, *Vice President—Directories*
Kevin Hayes, *Operations Manager*
Leif Gruenberg, *Development Manager*
David P. Bianco, *Marketing Consultant*

* * *

Peter E. Ruffner, *Publisher*

Frederick G. Ruffner, Jr., *Chairman*

ISBN 0-7808-0646-8

Library of Congress Cataloging-in-Publication Data

Communications security sourcebook : basic information for general readers about cell phone and wireless communication security, telephone company security issues, telephone slamming and cramming, long distance telephone scams, telemarketing fraud and other nuisances, wiretapping and eavesdropping, and more; along with glossaries of related terms and resources for further information / edited by Wilma R. Caldwell.-- 1st ed.
p. cm. -- (Security reference series)
Includes index.
ISBN 0-7808-0646-8
1. Telecommunication--Security measures. I. Caldwell, Wilma. II. Series.

TK5102.85.C68 2003
384.6'4'0289--dc21

2003051705

This book is printed on acid-free paper meeting the ANSI Z39.48 Standard. The infinity symbol that appears above indicates that the paper in this book meets that standard.

Printed in the United States

Table of Contents

Part II: Security Issues and Your Telephone Company

Part III: Long Distance Telephone Scams

Part IV: Telemarketing Fraud and Other Nuisances

Part V: Wiretapping and Electronic Eavesdropping

Part VI: Additional Help and Information

Preface

About This Book

Today's communications market has more options for consumers than ever before. According to the Cellular Telecommunications Industry Association, there are currently some 6.4 million cellular phones in operation. However, security concerns associated with advanced telecommunications technologies such as fraud, theft, privacy invasion, and government surveillance are rising.

The *Communications Security Sourcebook* provides basic consumer information about wireless communication security, telephone security, and information privacy, including wiretapping and eavesdropping, Caller ID, location tracking, long distance scams, telemarketing fraud, slamming and cramming, privacy issues with telephone companies, government surveillance issues, cell phone fraud and theft, telephone harassment and crank calls, wireless payment technology risks, and more. It also includes statistics, glossaries of related terms, and resources for further information.

How to Use This Book

This book is divided into parts and chapters. Parts focus on broad areas of interest. Chapters are devoted to single topics within a part.

Part I: Cell Phone and Wireless Security discusses cell phone theft and fraud, security issues related to smartphones, cell phone and other wireless device viruses, and wireless phone radiation risks.

Part II: Security Issues and Your Telephone Company explores the issues of slamming, cramming, Caller ID blocking, and how to avoid cellular phone surcharges, pre-paid phone card problems, and inappropriate use of your private information by your telephone company.

Part III: Long Distance Telephone Scams provides information on how long distance scams work and describes how to avoid specific telephone scams.

Part IV: Telemarketing Fraud and Other Nuisances informs readers on how to detect and avoid various telemarketing scams and how to deal with telephone harassment. It also discusses "do-not-call" lists, predictive dialing, and unwanted faxes.

Part V: Wiretapping and Electronic Eavesdropping provides information on the detection and prevention of eavesdropping. It reviews the security vulnerabilities of equipment such as answering machines, faxes, voice mail, wireless intercoms, baby monitors, cordless phones, and cellular phones. It also discusses roving wiretaps, aircracking, and issues related to governmental electronic surveillance.

Part VI: Additional Help and Information includes glossaries of telecommunications security terms, cellular and wireless terms, and wireless telephone fraud tips. It also includes directories of resources for further information.

Bibliographic Note

This volume contains documents and excerpts from publications issued by the following U.S. government agencies: Counterintelligence Training Academy, Department of Justice, Department of the Navy, Federal Bureau of Investigation (FBI), Federal Communications Commission (FCC), Federal Trade Commission (FTC), and the Food and Drug Administration (FDA).

In addition, this volume contains copyrighted documents from the following organizations, individuals, and publications: American Civil Liberties Union (ACLU); The Associated Press; Bennett Kobb; Beth Givens; Branden Shortt; Carmen Nobel; Cellphones.ca; Civil Rights Forum on Communications Policy; Cornell University; Dan Caterinicchia; Deborah Méndez-Wilson; Dennis Fisher; Electronic Frontier Foundation; Federal Computer Week; Graham Cluley; How Stuff Works, Inc.; iEntry, Inc.; Jim Krane; Jourdan Zayles; Lee Ann Obringer; Lee Tien; The Motley Fool; National Consumers League;

New Mexico Attorney General's Office; Ohio University Police Department; Patricia Madrid; Privacy Foundation; Privacy Rights Clearinghouse; SC Magazine; Simon Williams; Throop Wilder; Wireless Week; Ziff Davis Media, Inc.

Full citation information is provided on the first page of each chapter. Every effort has been made to secure all necessary rights to reprint the copyrighted material. If any omissions have been made, please contact Omnigraphics to make corrections for future editions.

Acknowledgements

Special thanks to Beth Givens at the Privacy Rights Clearinghouse and SC Magazine Online for going above and beyond to help meet deadlines and update material. The editor also wishes to thank Liz Barbour for her advice and assistance whenever needed.

Part One

Cell Phone and Wireless Security

Chapter 1

Introduction to Wireless Communications Security

Wireless phones are very popular, and the number of people who use them is steadily growing. There are 140 million subscribers in the U.S. But even though wireless devices have many advantages, privacy is not one of them.

Depending on the kind of phone you use, others can listen to calls you make. Pagers can also be intercepted. And if your computer is connected to a wireless network, the data you transmit to other computers and printers might not be secure.

It pays to be aware of the privacy and fraud implications of using wireless devices. A few simple precautions will enable you to detect and prevent fraud as well as to safeguard the privacy of your communications.

A word about terminology: This guide uses the terms "analog" and "digital" when describing wireless communications. **Analog** cellular services have been available for 25 years and are now available across 95% of the U.S. They send a voice through the air using a continuous radio wave. **Digital** services, available since 1995, convert the signal

---

"Fact Sheet 2: Wireless Communications Privacy: Wireless Communications: Voice and Data Privacy," Privacy Rights Clearinghouse, http://www.privacyrights.org/fs/fs2-wire.htm. Reprinted with permission from the Privacy Rights Clearinghouse, a non-profit consumer advocacy and information program located in San Diego, CA. © 2002. Contact the Privacy Rights Clearinghouse at 3100-5th Ave., Suite B, San Diego, CA 92103, (619) 298-3396 (voice), (619) 298-5681 (fax), prc@privacyrights.org (E-mail). For the most recent version of this fact sheet, or additional information, visit www.privacyrights.org.

into the ones and zeros of computer code. In contrast to analog signals which are continuous, digital transmissions are sent as varying pulses of electricity. Digital services do not yet have as wide geographic coverage as analog. But calls are generally clearer and more secure.

Cordless Telephones

Cordless phones operate like mini-radio stations. They send radio signals from the base unit to the handset and from the handset back to the base. These signals can travel as far as a mile from the phone's location.

Can Other People Listen to My Cordless Phone Conversations?

Yes, depending on the kind of phone you use. In most cases, your cordless phone conversations are probably overheard only briefly and accidentally. But there are people who make it a hobby to listen to cordless phone calls using radio scanners. These devices pick up the full range of wireless transmissions from emergency and law enforcement agencies, aircraft, mobile systems, weather reports, utilities maintenance services, among others. Signals from analog cordless phones can also be picked up by other devices including baby monitors, some walkie-talkies, and other cordless phones.

Newer digital cordless phones have better security, but cheaper or older phones have few if any security features. Anyone using a radio scanner can eavesdrop on older analog cordless phone calls, even if the phone has multiple channels.

What Privacy Features Should I Look for in a Cordless Phone?

When you shop for a new cordless phone, ask the sales clerk for an explanation of the privacy and security features. Read the product descriptions on the box, and visit the manufacturer's web site to obtain more information.

Cordless phones that operate on the higher frequencies (900 MHz or 2.4 GHz, pronounced "megahertz" and "gigahertz") are more secure, especially if they use digital spread spectrum technology and scramble the signal. But don't get a false sense of security that your conversations are totally immune from monitoring. Skilled hobbyists and determined professionals can monitor just about anything.

The fact that laws prohibit eavesdropping (discussed below) is rarely a deterrent. Unless the eavesdropper reveals details of the monitored conversations to you, it's virtually impossible to know if others are listening.

Since others can listen to cordless phone conversations, you should avoid discussing financial or other sensitive personal information. If you buy something over a cordless phone and give your credit card number and expiration date, you might end up the victim of credit card fraud.

Experts advise that the safest cordless phones operate at the higher frequency of 2.4 GHz. Digital models that use spread spectrum technology (SST) offer the best security. This feature breaks apart the voice signal and spreads it over several channels during transmission, making it difficult to capture. Because 2.4 GHz SST phones operate at higher power, they have another advantage, increased range.

Another security feature to look for is digital security codes. Both the handset and the base must have the same code in order to communicate. Look for phones that randomly assign a new digital code every time the handset is returned to the base.

Security codes do not prevent monitoring by radio scanners. But they do keep people nearby with similar handsets from attaching to your phone line to make their own calls and driving up your long distance bill. If your phone does not automatically change the security code for you after each use, remember to change it yourself. Do not use the security code set by the factory. Professional eavesdroppers know to search for those codes.

Don't be confused into thinking that just because your cordless phone has many channels it is more secure. However, if the phone automatically changes the frequency during communications, called channel hopping, it does provide more security by making it difficult for the eavesdropper to follow the call from one channel (frequency) to the next.

Beware of so-called security features that simply distort the analog signal. They make eavesdropping difficult but not impossible.

High-tech cordless phones are more expensive. If your budget is limited and you are not able to purchase a phone with these security features, remember to use a standard wired phone for all sensitive communications, including financial transactions. Be sure both you and the person you are talking to are on standard phones.

Special note about high-risk communications: If you have a high-profile occupation (entertainer, politician, corporate executive, high-ranking government official), if you're involved in a high-stakes lawsuit, if you are active in controversial political, religious, or social activities, or if you are a victim of stalking or domestic violence, you are more vulnerable to cordless phone eavesdropping. In fact, all of your electronic communications, whether wireless or wired, could be at risk. It is beyond the scope of this guide to suggest security strategies in these situations. Professional services are available that provide advice and technical assistance on securing high-risk communications.

Other Wireless Devices with Privacy Risks

Are There Other Gadgets or Services That May Be Broadcasting My Conversations?

Home Intercom Systems

Baby monitors, children's walkie-talkies and some home intercom systems may be overheard in the vicinity of the home in the same manner as cordless phones. Many operate on common radio frequencies that can be picked up by radio scanners, cordless phones, and other baby monitors nearby. If you are concerned about being overheard on one of these devices, turn it off when it is not in use. Consider purchasing a "wired" unit instead.

Speakerphones

If your standard wired phone has the speakerphone feature, be aware that some models may emit weak radio signals from the microphone even when the phone's handset is on-hook, (that is, hung-up, inactive). For short distances, a sensitive receiver may be able to pick up room noise in the vicinity of the speakerphone.

Wireless Microphones

Radio scanners can intercept wireless microphones used at conferences, in churches, by entertainers, sports referees, and others. Fast-food employees at drive-through restaurants use wireless systems to transmit order information. Their communications can also be received by scanners in the vicinity. Scanners can also pick up conversations on some walkie-talkies.

Wireless Cameras

Wireless videocameras have been installed in thousands of homes and businesses in recent years. The camera sends a signal to a receiver so it can be viewed on a computer or TV. These systems are advertised as home security systems, but they are far from secure. While they are inexpensive and relatively easy to install, they are also easy to monitor by voyeurs nearby who are using the same devices.

Images can be picked up as far as 300 yards from the source, depending on the strength of the signal and the sensitivity of the receiver. Before purchasing a wireless videocamera system, ask yourself if you want to be vulnerable to electronic peeping toms. Research the security features of such systems thoroughly. You might want to wait until the marketplace provides wireless video systems with stronger security features at an affordable price.

Air-to-ground Phone Services

Conversations on the phone services offered on commercial airlines are easily intercepted by standard radio scanners. They are a favorite target of hobbyists.

Cellular Telephones

Cellular phones send radio signals to low-power transmitters located within "cells" that range from the size of a building to 20 miles across. As you travel from cell to cell, the signal carrying your voice is transferred to the nearest transmitter.

Can Others Listen to My Cellular Phone Calls?

Yes, depending on the phone system's technical features. Cellular phone calls usually are not picked up by electronic devices such as radios and baby monitors. But analog cell phone transmissions can be received by radio scanners, particularly older model scanners and those that have been illegally altered to pick up analog cell phone communications.

With advances in digital technology, wireless voice communications are much more difficult to intercept than analog phones. The digital signal that is received by a standard radio scanner is undecipherable and sounds like the noise made by a modem or fax machine when transmitting over phone lines. Law enforcement-grade scanners can monitor digital communications, but these are expensive and generally not available on the open marketplace

Many digital phone models are dual- and even tri-mode. They enable the user to switch to analog mode when digital services are not available. Remember that in analog mode, conversations can be monitored on standard radio scanners.

What Technical Features Should I Look for in Cell Phones to Protect My Privacy?

As with cordless phones, digital cell phones are more secure than analog phones by default. Phone conversations on digital phones cannot be picked up by the kinds of radio scanners used by casual hobbyists. Nonetheless, there are features you should consider regarding digital phone security.

Digital communications that are encrypted provide the highest security. Several digital technologies are available in the U.S., primarily CDMA,* TDMA,* and GSM.* But few carriers here encrypt digital transmissions, in contrast to Europe.

In the U.S., CDMA systems that use the spread spectrum (SST) technology provide very strong security, difficult to intercept except by law enforcement and skilled technicians. The next generation of GSM systems, 3G, will also use SST and according to experts will have strong security.

(*Terminology: These abbreviations designate the technical interface used by the carrier. CDMA stands for code division multiple access. TDMA is time division multiple access. GSM means Global System for Mobile Communications. GSM is more common in Europe, but some U.S. carriers are converting to it. Two other interfaces are ESMR, or Enhanced Specialized Mobile Radio, and iDEN, which is based on TDMA.)

Another term associated with digital phones is PCS, personal communications services. While "standard" digital cellular operates at 800 MHZ, PCS operates at 1900 MHZ. The types of services available on PCS are much the same as digital cellular. To learn more, read "How Wireless Phone Technology Works," at www.ctia.org, the web site of the industry organization Cellular Telecommunications and Internet Association.

Are There Other Privacy Risks of Cell Phone Use?

Some cell phone models can be turned into microphones and used to eavesdrop on conversations in the vicinity. This is why some businesses and government agencies prohibit cell phones in areas where sensitive discussions are held.

And don't forget (although many cell phone users do): Your side of the conversation can be heard when you talk on your cell phone in crowded public places like restaurants, airports, malls, public transportation, and busy city streets. If you don't want others to listen to your personal conversations, be discreet and speak softly. Better yet, move out of earshot of others or save those conversations for the privacy of your car, home, or office.

What Are the Privacy Implications of Location-Tracking Features?

By 2005 the Federal Communications Commission has mandated that the majority of wireless providers be able to locate 911 calls within about 100 feet so that emergency services can find callers using cellular phones. This feature is called E-911. (www.fcc.gov/911/enhanced) Carriers can either provide the location information that resides in the cellular network (triangulation of location based on the distance of the cell phone's signal to nearby cellular towers), or they can rely on satellite data from global positioning system (GPS) chips embedded in the handsets of their customers.

The requirement that cell phones be embedded with location-tracking technology has spawned a new industry—location-based services such as targeted advertising. Here's how it is expected to work. As your car approaches a freeway exit where a restaurant features your favorite food, you could receive a text message on your phone or handheld device with a special offer. Or as you walk past a coffee house, your phone could receive an ad offering you a discount on a double latté.

While some might welcome this form of advertising, others are concerned about the privacy implications of location-based advertising. After all, in order to send you such ads, the service must know something about your interests as well as your specific location. If location records were kept over time, an in-depth profile could be compiled for both marketing and surveillance purposes.

One of the first location-based services to enter the marketplace is AT&T's Find Friends. It enables users to pinpoint another AT&T user's cell phone location, depending on how close that person is to the nearest cellular tower. Find Friends can also locate nearby businesses and invite another user of the service to meet there by sending them a text message. Find Friends uses a feature similar to "buddy lists," borrowed from Internet instant messaging systems. The user invites others to be on the list and therefore to be "locatable" in order to receive text messages on their cell phones. Buddies can key instructions into the phone to become "invisible" when they do not want to be located. Other wireless carriers are expected to develop similar services.

The wireless industry is aware of consumers' privacy concerns and has been working to develop consent-based guidelines for the development of wireless advertising. In August 2002, the Federal Communications Commission (FCC) turned down the wireless industry's request to adopt location information privacy rules. The proposed rules were based on the privacy principles of notice, consent, security, and integrity of consumer data.

Because of the federal government's reluctance to regulate location-based wireless services, consumers must carefully research the privacy implications of these services before subscribing. Individuals are encouraged to only subscribe to services that offer maximum user control. Not only must users be able to turn off location-tracking features, industry must ensure that the wireless devices come out of the box with location tracking turned off, with the exception of E-911 calls. Further, one's "locatability" and the receipt of targeted ads should be subject to an "opt-in," requiring the user's affirmative consent.

Be sure to carefully read the privacy policy of any wireless service you are considering, usually available on the company's web site and on product brochures. Pay attention to how the service captures and stores data, and what it says about the retention of customer and location data. To maximize privacy protection, avoid services that store location data.

Are There Fraud Risks Involved with Using a Cellular Telephone?

There are two types of fraud risks—cell phone "cloning" and subscription fraud, also known as identity theft. Cloning has declined dramatically in recent years, while subscription fraud is increasing.

In the mid-1990s, cloning of cell phone electronic serial numbers (ESN) was rampant. Cell phone companies lost several hundred million dollars each year to cloning. The ESN is a unique serial number programmed into the cellular phone by the manufacturer. The ESN and the Mobile Identification Number (MIN) are used to identify a subscriber. One way the ESN is cloned is by capturing the ESN-MIN over the airwaves. The ESN-MIN is then reprogrammed into a computer chip of another cellular telephone. The phone calls made by the cloned phone are listed on the monthly bill of the person whose phone was cloned.

Cell phone cloning has declined significantly in recent years. The industry developed authentication features that have greatly reduced cell phone cloning, although some still occurs.

Today, the cell phone industry is battling subscription fraud, also known as identity theft. An imposter, armed with someone else's Social Security number, applies for cell phone service in that person's name but the imposter's address. As with other forms of credit-related identity theft, the imposter fails to pay the monthly phone bills and phone service is eventually cut off.

When the phone company or a debt collection company attempts to locate the debtor, it finds, instead, the victim who is unaware of the fraud. That person is then saddled with the long, laborious process of settling the matter with the phone company and repairing his or her credit report. The Federal Trade Commission reports that phone/utilities fraud is the second most common form of identity theft following credit fraud, half of which is wireless subscription fraud. (FTC Sentinel, www.consumer.gov/sentinel)

What Can Be Done to Prevent Cellular Telephone Fraud?

You can prevent cloning, or at least detect it early, by taking these steps:

- Check with the cellular phone company to find out what anti-fraud features they have. Make sure the service you select uses authentication technology to prevent cloning.
- Always use the phone's lock feature when you are not using the phone.
- Do not leave your phone unattended, or in an unattended car. If you must leave it in your vehicle, lock the phone out of sight and use the phone's lock code.
- Keep documents containing your phone's ESN in a safe place.
- Check your cellular phone bills thoroughly each month. Look for phone calls you did not make and report them immediately to the phone carrier.
- If you receive frequent wrong numbers or hang-ups, these could be an indication that your phone has been cloned. Report these to the phone carrier right away.
- Ask the phone carrier to eliminate overseas toll calls or North America toll calls if you do not intend to make long distance calls.
- Report a stolen cellular telephone immediately to the cellular telephone carrier.

Consumers usually learn about cellular telephone cloning when they receive their bill. However, standard industry practice is to not charge consumers for cloned calls. If you fall victim to cloning, contact your cellular telephone provider immediately. If you are having a problem with your service provider, file a complaint with the Federal Communications Commission. See the Resources section at the end of this guide for more information.

Subscription fraud is another matter. Your existing cell phone is not the target of fraud as in cloning. Rather, an imposter has established a new phone account in your name, with the monthly bills sent to their address, not yours. You usually don't find out about it until the bills are long past due and a debt collector tracks you down.

Early detection is the key to minimizing the aggravation of subscription fraud. Be sure to check your credit report at least once a year. If someone else has a cell phone in your name, you will notice an "inquiry" from the phone company on your credit report. And if the account has gone to collection, it is likely to be noted on the credit report. You will not be responsible for paying the imposter's bills, but you will need to take the necessary steps to remove the fraudulent account and/or inquiry from your credit report. (Read the PRC's guide, "Identity Theft: What to Do if It Happens to You," at www.privacyrights.org/identity.htm.)

Are There Laws That Prohibit Cellular Telephone Fraud?

Yes. Federal law makes it a crime to knowingly and intentionally use cellular telephones that are altered, to allow unauthorized use of such services. (18 USC 1029) Penalties for violating this law include imprisonment and/or a fine. The Secret Service is the agency authorized by this law to investigate cellular phone fraud.

In California, it is a crime to intentionally avoid a telephone charge by the fraudulent use of false, altered or stolen identification. (California Penal Code 502.7) In addition, it is against the law to use a telecommunications device with the intent to avoid payment for service. Penalties include imprisonment and/or a fine. (California Penal Code 502.8)

The California Public Utilities Commission requires cellular telephone service providers to give their subscribers a notice that warns them of problems associated with fraud and provide them with information on ways to protect against fraud. (California Public Utilities Code 2892.3)

Subscription fraud is also a crime. The federal law is the Identity Theft and Assumption Deterrence Act (18 USC 1028). Most states have also criminalized identity theft. The Federal Trade Commission provides information about these laws and how to recover from identity theft. The FTC's identity theft clearinghouse can be contacted at (877) IDTHEFT, and its web site is www.consumer.gov/idtheft. The Privacy Rights Clearinghouse (www.privacyrights.org) and the Identity Theft Resource Center (www.idtheftcenter.org) offer additional information.

Pagers and Other Messaging Devices

There are several types of pagers on the market: tone-only pagers (which are outmoded and rarely used any more), numeric, alphanumeric, and two-way pagers. Pagers can be either purchased or rented. The monthly fees can be significantly less than cellular or standard phone services. The costs depend on the type of pager and services the subscriber wants to receive.

Can Pager Communications Be Monitored? What about Other Text Messaging Systems?

Pager messages are not immune to monitoring. Pager networks are generally not encrypted. They transmit in the frequencies that can be monitored by radio scanners, although messages cannot be deciphered without special equipment attached to the scanner. Hackers trade tips on web sites on how to intercept pager messages. Law enforcement-grade devices are available that pick up pager communications.

The odds of your pager messages being intercepted and deciphered are probably low, especially given the cryptic nature of these messages. But individuals who engage in high-risk communications, as discussed in the cordless phone section of this guide, should take appropriate precautions.

Text messaging via cell phones and "handhelds" is much more advanced in Europe and Asia than the U.S., although it is rapidly growing here. (Handhelds are small, highly portable computer/communications devices such as personal digital assistants, or PDAs.) The jury is still out on whether or not short message services (SMS) on cell phones and other wireless devices can be intercepted. But there have been reports of the transmission of fake text messages, called "SMS spoofing."

Be sure to thoroughly research the security features of any messaging system that you use. Experts advise that you install security features on your handhelds to encrypt data that you transmit and to prevent others from accessing your data if the device is lost or stolen.

Laws Regarding Wireless Eavesdropping

Is It Legal to Intercept Other People's Cordless or Cellular Phone Calls?

The Federal Communications Commission (www.fcc.gov) ruled that as of April 1994 no radio scanners may be manufactured or imported into the U.S. that can pick up frequencies used by cellular telephones, or that can be readily altered to receive such frequencies. (47 CFR Part 15.37(f)) The law rarely deters the determined eavesdropper, however.

Another federal law, the Counterfeit Access Device Law, was amended to make it illegal to use a radio scanner "knowingly and with the intent to defraud" to eavesdrop on wire or electronic communication. (18 USC 1029) Penalties for the intentional interception of cordless and cellular telephone calls range from fines to imprisonment depending on the circumstances. (18 USC 2511, 2701)

There are exceptions in electronic eavesdropping laws for law enforcement monitoring. The Communications Assistance for Law Enforcement Act of 1994 (CALEA) requires telecommunications carriers to ensure that their equipment, facilities, and services are able to comply with authorized electronic surveillance by law enforcement. (www.fcc.gov/calea) The FBI's (Federal Bureau of Investigation's) CALEA web site is www.askcalea.com.

Under California law it is illegal to intentionally record or maliciously intercept telephone conversations without the consent of *all* parties. This includes cordless and cellular calls. (California Penal Code 632.5-632.7) To violate the law, the interception of your cordless or cellular phone conversations must be done with malicious intent.

So, if your neighbor accidentally hears your cordless phone conversation on a radio scanner, it's probably not illegal. But unless the eavesdropper discloses what he or she has overheard, you have no way of knowing your conversation has been monitored. Even though an eavesdropper would be violating the law, it's not likely that you or anyone else will detect it.

There are some exceptions to California's all-party consent law. A judge can authorize the interception of an electronic cellular telephone communication in investigations involving specified crimes. (California Penal Code 629.50-629.98) California Penal Code section 633.5 states that if someone is threatening another person with extortion, kidnaping, bribery, or any other felony involving violence, the calls may be recorded by the person being threatened. Under special limited circumstances, phone company employees may monitor calls.

Laws in other states vary. In fact, the weaker standard of *one-party consent* is law in a majority of states. For a directory of the wiretapping and eavesdropping laws in the 50 states, visit the web site of the Reporters Committee for Freedom of the Press (www.rcfp.org/taping). The National Conference of State Legislatures provides a chart of federal and state electronic surveillance laws, www.ncsl.org/programs/lis/CIP/surveillance.htm.

Are There Laws Related to the Privacy of Pagers?

Federal law prohibits anyone from intercepting messages sent to display pagers (numeric and alphanumeric) and to tone-and-voice pagers. Tone-only pagers are exempt from this provision. (Electronic Communications Privacy Act, 18 USC 2510)

Law enforcement must obtain a court order in order intercept your display or tone-and-voice pager. But under the USA PATRIOT Act, enacted in 2001 following the September 11 terrorist attacks, the standards for obtaining court ordered warrants have been loosened.

In California, a judge can authorize the interception of an electronic digital pager by law enforcement in investigations involving certain specified offenses. (California Penal Code 629.50)

Can Telemarketers Contact Wireless Phones, Pagers, and Other Text Devices?

Under the federal Telephone Consumer Protection Act, it is against the law to use autodialers or prerecorded messages to call numbers assigned to pagers, cellular or other radio common carrier services except in emergencies or when the person called has previously communicated their consent. (47 USC 227)

But the law fails to specifically prohibit “live” telemarketing calls to cell phones. Telemarketers claim that they do not target cell phones with solicitations, but it can happen, especially if a wireline phone number is inadvertently assigned to a cell phone. Aside from the privacy and annoyance factors of receiving junk calls on cell phones, there is the further aggravation of having to pay for those calls. (Cell phone users generally pay for both the outgoing and incoming calls.)

As wireless text messaging systems become more widespread, it is only a matter of time before “spam”—unsolicited electronic bulk advertising—becomes a problem for wireless consumers. A bill introduced in Congress in 2001, the Wireless Telephone Spam Protection Act (H.R. 113), would make it illegal to transmit unsolicited ads to wireless devices, including cell phones, pagers, and PDAs enabled to receive wireless e-mail.

A new law in California, effective 2003, prohibits the transmission of text message advertisement to cellular phones or pagers equipped with short message capability. The law has exceptions if the company has an existing relationship with the subscriber or if it gives customers the option to not receive text messages. (California Business and Professions Code 17538.41)

Wireless Data Networks

An increasing number of households and businesses are establishing wireless networks to link multiple computers, printers, and other devices. A wireless network offers the significant advantage of enabling you to build a computer network without stringing wires. Unfortunately, these systems usually come out of the box with the security features turned off. This makes the network easy to set up, but also easy to break into. Most wireless networks use the 802.11 protocol, also known as Wi-Fi.

What Are the Security Risks of Using Wireless Data Networks?

Wireless networks have spawned a new past-time among hobbyists and corporate spies called war-driving. The data voyeur drives around a neighborhood or office district using a laptop and free software to locate unsecured wireless networks in the vicinity, usually within 100 yards of the source. The laptop captures the data that is transmitted to and from the network's computers and printers. The data could include anything from one's household finances to business secrets.

Wireless network units are equipped with security features, but they are usually disabled, requiring you to turn them on when you install the system. Not only can data be stolen, altered, or destroyed, but programs and even extra computers can be added to the unsecured network without your knowledge. This risk is highest in densely populated neighborhoods and office building complexes.

Remember, wireless data networks are in their infancy. To ensure that your system is secure, review your user's manuals and web resources for information on wireless security. The Home PC Firewall Guide provides access to independent, third-party reviews of Internet security products, including wireless computer networks and PDAs, on the web at www.firewallguide.com/index.htm. Another useful guide can be found on the web at www.practicallynetworked.com/support/wireless_secure.htm.

Resources for More Information

[See the "Additional Help and Information" section of this *Sourcebook* for further information on resources.]

Chapter 2

Cell Phone Theft

Everywhere you go, there seem to be people talking on cell phones. Many people carry a cell phone for business and other personal calls. We may not give it a second thought to leave the cell phone in the car while we are at a movie. Those who use a cell phone for emergencies may have to take a moment to remember where the phone was left. I would like to suggest to you that you need to keep track of your cell phone, or you may be in for quite a shock.

If your cell phone is stolen, you could be financially responsible for the calls made on the phone. My office has received complaints from individuals who are being charged for thousands of dollars in calls made after their cell phone was stolen. If you find yourself in such a situation, you could be in for a prolonged dispute, as well as possible credit problems.

Many people think that their liability for cell phone charges is similar to a credit card theft. But that is not that case. If you read your cell phone contract, you are likely to find no liability limit because cell phone providers are not required to limit your liability.

I would like to suggest the following steps for your protection:

"Topic: Cell Phones," New Mexico Attorney General, http://www.ago.state.nm.us/Alert/Consumer_Awareness_Topics/CellPhones.htm. Additional consumer information is available on the website of the New Mexico Office of the Attorney General at www.ago.state.nm.us. Residents of New Mexico may contact the Attorney General's office for assistance at P.O. Drawer 1508, Santa Fe, NM 87504-1508, (800) 678-1508.

- Know where your cell phone is at all times. If days go by before you notice your phone was stolen, you could be faced with paying a sizable cell phone bill.
- Read your cell phone contract carefully. The contract may contain specific instructions to be followed in the event your cell phone is stolen.
- If your phone is stolen, you should notify the police so there is a written record of the theft and notify the cell phone provider as quickly as possible.
- When you talk to your cell phone provider about the theft, make certain you ask the name of the individual you are speaking with and that person's position. You might need this information later if the company maintains that the theft was not reported.

Chapter 3

Cell Phone Fraud: Unauthorized Use of Your Account

Cellular fraud (cell fraud) is defined as the unauthorized use, tampering, or manipulation of a cellular phone or service. At one time, cloning of cellular phones accounted for a large portion of cell fraud. As a result, the Wireless Telephone Protection Act of 1998 expanded prior law to criminalize the use, possession, manufacture or sale of cloning hardware or software. Currently, the primary type of cell fraud is subscriber fraud. The cellular industry estimates that carriers lose more than $150 million per year due to subscriber fraud.

What Is Subscriber Fraud?

Subscriber fraud occurs when someone signs up for service with fraudulently-obtained customer information or false identification. Lawbreakers obtain your personal information and use it to set up a cell phone account in your name.

Resolving subscriber fraud could develop into a long and difficult process for victims. It may take time to discover that subscriber fraud has occurred and an even longer time to prove that you did not incur the debts. Call your carrier if you think you have been a victim of subscriber fraud.

"FCC Consumer Alert: Cell Phone Fraud," Federal Communications Commission (FCC), http://www.fcc.gov/cgb/consumerfacts/cellphonefraud.html, reviewed/updated on February 12, 2002.

What Is Cell Phone Cloning Fraud?

Every cell phone is supposed to have a unique factory-set electronic serial number (ESN) and telephone number (MIN). A cloned cell phone is one that has been reprogrammed to transmit the ESN and MIN belonging to another (legitimate) cell phone. Unscrupulous people can obtain valid ESN/MIN combinations by illegally monitoring the radio wave transmissions from the cell phones of legitimate subscribers. After cloning, both the legitimate and the fraudulent cell phones have the same ESN/MIN combination and cellular systems cannot distinguish the cloned cell phone from the legitimate one. The legitimate phone user then gets billed for the cloned phone's calls. Call your carrier if you think you have been a victim of cloning fraud.

Summary

Remember, to prevent subscriber fraud, make sure that your personal information is kept private when purchasing anything in a store or on the Internet. Protecting your personal information is your responsibility. For cell phone cloning fraud, the cellular equipment manufacturing industry has deployed authentication systems that have proven to be a very effective countermeasure to cloning. Call your cellular phone carrier for more information.

Chapter 4

Frequently Asked Questions about Wireless Telephone Fraud

Q: How Are Fraudulent Calls Made?

There are different techniques, but in essence the criminal takes apart a wireless telephone and reprograms it with a counterfeit account code (ESN/MIN pair), which tricks a wireless system into sending the bill elsewhere.

Q: Does the Customer Get Stuck with the Bill If Someone Fraudulently Uses His or Her Account?

No. It has been the policy of wireless carriers to remove fraudulent charges from the accounts of customers. However, it's important to remember that wireless fraud is not a victimless crime. It adds to the cost of doing business, and legitimate customers are inconvenienced.

Q: How Big Is the Wireless Telecommunications Fraud Problem?

In 1997, the wireless industry lost $434 million due to fraud. The losses for 1997 represent 1.4 percent of total industry revenues. It is

"Wireless Telephone Fraud FAQ's: Wireless Telephone Fraud: Frequently Asked Questions," Cellular Telecommunications and Internet Association (CTIA), http://www.wow-com.com/industry/tech/security/articles.cfm?ID=309.

estimated that losses for 1998 will be 182 million or .05% of industry revenues.

Q: What Is the Industry Doing to Stop Fraud?

Wireless carriers are waging a high-tech war against high-tech criminals. They are attacking on several fronts. The industry supports field investigations of criminal operations; education programs for law enforcement and carrier personnel; strengthening legal sanction; and research into new technological solutions including: Radio Frequency Fingerprinting, Personal Identification numbers, Roamer Verification, Reinstatement Profile and Authentication.

Q: Is the Fraudulent Activity Only in the Big Markets?

No. While the bulk of the problem is in bigger cities such as New York , Miami, and Los Angeles, wireless fraud can take place anywhere.

Q: Is It Illegal to Tamper with a Wireless Phone?

On October 25, 1994, H.R. 4922 was signed into law by President Clinton. Amendments to Section 1029 now include the fraudulent alteration of telecommunications instruments and equipment. Punishment includes fines of up to $50,000 and 15 years imprisonment. Additionally, the rules and regulations of the Federal Communications Commission prohibit tampering with and/or altering the Electronic Serial Number (ESN) inside a wireless telephone. Every wireless phone must have a unique ESN and no two phones may have or emit the same ESN, according to FCC rules. On April 24, 1998 President Clinton signed Senate Bill 493 making it a violation of Title 18, Section 1029 to knowingly use, produce, or traffic in or have control of or custody of, or possess hardware or software to counterfeit wireless phones.

Q: Who Helps the Industry on the Law Enforcement Side?

Local, county, state and federal police agencies, depending on the case and location. On the federal side, most federal assistance comes from the United States Secret Service in the U.S. and the Royal Canadian Mounted Police in Canada because wireless fraud involves electronic counterfeiting. Other federal agencies now pursuing these cases include the Federal Bureau of Investigation (FBI), Drug Enforcement Administration (DEA) and U.S. Customs.

Chapter 5

Wireless Telephone Fraud Consumer Tips

All wireless phone users should know that fraud can be prevented at the consumer level, both in terms of prevention and detection.

Consumers Can Help Prevent Fraud by:

1. Locking phones or removing handsets and wireless antennas (to avoid drawing attention to the vehicle) every time a vehicle is left with someone, like a parking lot attendant or mechanic.
2. Protecting sensitive documents such as subscriber agreements, which include electronic serial numbers.
3. Immediately reporting a stolen phone to the wireless phone carrier.
4. Not leaving their phone in an unattended car in an isolated or questionable area or parking lot for an extended period of time; locking the phone out of sight; and using the lock code.

"Wireless Telephone Fraud Consumer Tips," Cellular Telecommunications and Internet Association, http://www.wow-com.com/industry/tech/security/articles.cfm?ID=310.

Consumers Can Help Detect Fraud by:

1. Looking for unusual call activity on their monthly wireless phone bill.
2. Reporting frequent receipt of wrong numbers or hang ups on the wireless phone, which may indicate someone else is using their mobile number.
3. Asking the wireless provider to eliminate overseas toll or North American toll (long distance) dialing capabilities if the customer does not intend to call long distance.
4. If a wireless subscriber suspects fraud, he or she should immediately contact the wireless phone company.

Chapter 6

Third Generation (3G) Wireless: Does Converging Technology Mean Converging Fraud?

Mention imminent third generation (3G) mobile networks and thoughts will most likely turn to the radical new services they will deliver or the staggering sums that must be paid for related operating licenses.

The investment committed to 3G is immense. Understandably, network operators are devoting their best efforts to minimizing any elements that could influence the revenue flows necessary for redeeming this investment.

All the same, it's a safe bet that when the first 3G wireless systems open for business, the first incidents of 3G fraud will follow close behind. Forewarned is forearmed, which is why now is the time to examine likely scenarios and solutions. At this stage, a degree of speculation is inevitable, but we can at least identify some of the issues that will need to be addressed.

Is Fraud Inevitable?

The answer to this question is debatable, but it's a fact that people will always try to exploit weaknesses in any telecom network. Attractive rewards and a perceived low risk of detection make mobile communications particularly susceptible.

"3G: Does Converging Technology Mean Converging Fraud?," SC On-Line, SC Magazine, 161 Worcester Road, Suite 201, Framingham, MA 01701, December 2000.

Accepted estimates show that network operators of global systems for mobile communications (GSM) are suffering fraud losses of three to five percent of their organizations' annual revenue. At the same time, annual losses due to fraud are expected to cost the global telecom industry over US$30 billion by 2002.

There is an important lesson to be learned from the prepaid experience in the mobile phone market. It was hoped that prepaid charging would reduce companies' exposure to fraud but new forms have quickly developed around the various token and credit card top-up methods. The rapid growth of fraud in the prepaid market caught many network operators by surprise. It is essential, therefore, that operators don't make the same mistakes when launching their 3G services.

Nature of the Beast

As 3G will be built on the convergence of several core technologies, the new services will provide a natural meeting point for three broad categories of fraud, as explained below:

- Telecom Voice Fraud—generally, these encompass fairly 'low-tech' types of fraud that are exploited by large numbers of small-time operators;
- Data Fraud—this includes IP front-end fraud and hacking carried out by computer software hackers and code writers;
- Credit Card/Financial Services Fraud—conducted by fraudsters who set up accounts in their own name; this type provides unauthorized access to funds.

As an illustration, banking services offered over an IP link, via a mobile device, will provide rich pickings for the smart fraudster who is able to combine these three fraud types to exploit an illegal profit potential that 3G typically will provide.

Because 3G networks will be designed as channels for relatively high-value transactions, the risks of fraud will escalate significantly. This is a function of falling call charges combined with the introduction of myriad services across the network such as banking and retailing. It is vital that 3G operators take active steps to minimize their exposure by protecting known areas of vulnerability.

Nortel Networks Fraud Solutions (NNFS) has identified the three major issues of concern that are related to these known exposures.

Contractual Liability

This concerns the contractual relationship between the network operator, the service provider(s) and customers. The critical question is how to judge the financial value of the content of a 3G phone call and who carries the responsibility for that value at various points during a transaction. In short, when something goes wrong or fraud occurs, who pays for the loss?

Only for simple purchases would the loss be confined to the value of the failed transaction. If loss or theft of 'intangible' goods occurred, for example, in share dealing, the network operator would need to avoid liability when customers incurred heavy losses as a consequence of a 'hacked' transaction or of being unable to contact their online share service to buy or sell.

There are enormous security implications here. A stolen phone could give complete access to owners' PIN codes, credit facilities, share portfolio and banking details—indeed, to their total identities. In addition, network operators will need to consider the practicalities of how and to whom payment for purchased goods is made. Will they, for example, be added to the cost of a phone bill? These are complex areas requiring expert guidance from the legal profession.

Operational Implications for the Fraud Team

The launch of 3G services will create a dramatic shift from current practice in which every call generates a definable call data record (CDR) that allows charges to be easily determined and measured, according to the destination and duration of a call. The CDR is the key to fraud management, credit and debt management, revenue assurance and, to an increasing extent, the sales and marketing function.

With 3G, the value of calls will be determined by new and as yet undefined criteria that will be much more complex to predict, measure and cost. Variables, such as the amount of bandwidth used, the value of the transaction or the type of data downloaded, will all be factors to take into account. At the current time, the nature of 3G charging is as open to interpretation as are the predictions about which services will be most popular and profitable. Services will have different value potential, whether downloading data, graphics, video, music or playing games online. Furthermore, it will no longer be possible to determine call destination.

Whatever conventions are agreed for pricing 3G services, a sophisticated new model will be needed for assessing market value. This will not be possible to gauge by call records alone.

Technological Implications for Fraud Analysis

Success in combating fraud will depend on the criteria used to analyze caller profiles, which will be presented in many different ways. Analysts will need to understand the multiple aspects of constructing a 3G-caller profile to observe what is happening on their network before they can begin to identify potential types of fraud.

One certainty is that traditional rule-based solutions will be totally inadequate for dealing with 3G fraud. Together with sophisticated analytical tools for profiling callers, there will be a need for teams to cope with new frauds delivered both at the telephony platform (from hackers and phreakers) and at the fixed Internet protocol (IP) and IT infrastructures.

The latter will extend beyond the traditional telecom domain into the banking and commercial arenas, where a major threat will be from insiders with direct access to the data streams. Of the current fraud threat a significant proportion is believed to arise from internal activity, occurring because authorized (in other words, insider) users have the ability to get away with unauthorized acts undetected. Within an IP environment and across enterprise networks this situation is likely to become an increasing issue, putting pressure on network operators to maintain the highest standards of vigilance internally as well as externally.

Developing Solutions for 3G Fraud

While there is much uncertainty about the impact of 3G, telecom operators can draw reassurance from new techniques and solutions that are already at an advanced stage of development. NNFS, for instance, is committed to an evolutionary strategy that builds on its experience in fraud detection and management.

In the 3G context, the behavior anomaly analysis approach (based in part on neural networks) puts operators in an advantageous position to tackle fraud, as attention is focused on data access, not content. Being one step removed from data and IP packets, this approach allows the issue to be visualized more clearly than in the traditional rules-based system, which would be able to adapt much less efficiently, if at all.

NNFS is already using similar techniques for tackling IP networks with a number of its clients. The real issue is the ability to analyze the IP data packet service, where content is invisible and destination effectively unknown.

Learning Curve

Inevitably there will be a steep learning curve for network operators, service providers, hardware manufacturers and fraud management solution providers. All of these professionals will face the challenge of developing countermeasures for a threat that is not yet in existence.

Much can be learned from previous experience and the proven methods already developed by fraud management specialists. This past knowledge represents the best weapons for controlling future threats. To meet this need, the NNFS user forum, Club Cerebrus, convenes twice a year to ensure that our fraud management research and development program is conducted in close collaboration with customers, some of whom are the world's leading telecom operators.

Undoubtedly, as commercial strategies for 3G develop and the nature of 3G fraud becomes increasingly apparent, even better defenses will result.

—by Simon Williams

Simon Williams is vice president of sales and marketing for Nortel Networks Fraud Solutions (NNFS).

Chapter 7

Addressing Wireless Device Privacy and Security Concerns

Privacy Issues

There was widespread agreement among workshop participants that emerging wireless technologies raise not only many of the privacy issues encountered in the wired world, but new privacy issues as well. This section will discuss the privacy concerns raised by wireless technologies and some possible ways to address them.

Privacy Concerns

Panelists listed the following as the most important privacy concerns related to wireless technologies: collection of location information, tracking visits to wireless websites, and increased personal data collection.

Collection of Location Information

Panelists generally agreed that the generation and potential use of location-based information is one of the most significant privacy issues in the wireless space. Many panelists, representing both industry

---

Excerpted from "The Mobile Wireless Web, Data Services and Beyond: Emerging Technologies and Consumer Issues," Federal Trade Commission (FTC), http://www.ftc.gov/bcp/reports/wirelesssummary.pdf. This document is dated February 2002, and is an update of a public workshop held by the FTC on December 11 and 12, 2000. Readers are advised that there is more information in the full document, including footnotes and listings of participants.

and consumer groups, stated that location-based services raise concerns because the consumer's specific location can be tracked whenever the user's device is on, which could be a significant portion of the day. Panelists recognized that personally identifiable location information is extremely sensitive. A representative of the Center for Democracy and Technology stated that companies will be able to track location in a way that was never available before, and many consumers do not know about this technology. A privacy and security consultant stated that the collection of detailed location information provides opportunities for abuse of the information. In addition, an industry representative stated that even if the consumer consents to a specific service provider obtaining the location information, once the service provider has the information, others could obtain the data through a court order. Another industry representative expressed concern that certain companies may consider a consumer's location and location history to be information that should be available to any entity possessing the technology to capture it.

According to workshop participants, a typical cell phone that is turned on sends out signals every ten minutes and identifies the location of the nearest cell tower. In less populated areas, cell towers are located about every thirty miles, but in cities, cell towers are located about every two blocks. These signals can be used to determine a cell phone user's location, so that even without the precise auto-location technologies discussed above (e.g., GPS), a user's location can be pinpointed fairly precisely in some areas. For the most part, carriers are not currently archiving this location information; however, companies may develop new business models to archive and use such information in the near future. Moreover, GPS and the other auto-location technologies will soon generate location data that identifies the user's precise location at any given moment when the device is turned on.

As discussed at the workshop, the FCC has issued a set of rules, called the Enhanced 911 ("E911") rules, that require wireless carriers to collect precise location information in the near future in order to improve the delivery of emergency services. Today, although a 911 operator receives exact information about a caller's location upon receiving a landline call, that is not the case with all cellular calls to 911. For many cellular calls, the operator has to ask for the caller's location first, slowing down the needed emergency services. The problem is exacerbated when the caller does not know where he or she is, and thus the dialogue can take a significant amount of time.

The FCC's E911 rules require common carriers to adapt their wireless networks so that they automatically provide certain information to 911 call centers. The initial rules, which have already been implemented in some areas, require wireless carriers to provide the caller's telephone number and generalized location information—typically the location of the cellular tower nearest to the caller—to the call center when a consumer dials 911. The rules then require carriers to begin implementing precise auto-location technology for transmission to 911 call centers. Once these procedures are fully implemented, 911 operators will be able to identify a cellular caller's precise location immediately upon picking up the call.

An industry representative stated that, because carriers are required to collect precise location information for emergency purposes, it is likely that businesses will find commercial applications for the information once collected. To address privacy concerns arising from these potential commercial uses, a provision of the Wireless Communications and Public Safety Act of 1999 amended the Telecommunications Act to provide that carriers must obtain "express prior authorization" before releasing this location information to third parties. The FCC has not yet begun a rulemaking to provide further guidance on implementing this statutory provision.

The Wireless Communications and Public Safety Act of 1999 also provides that information collected about a consumer's location is to be treated as customer proprietary network information ("CPNI"), which is data about a customer's telephone service and usage that receives special legal protections under Section 222(c)(1) of the Telecommunications Act of 1996. The 1996 statute provides that carriers must obtain "the approval of the customer" before sharing CPNI with a third party (unless a specific statutory exception applies). In 1998, the FCC issued an implementing regulation that defined "approval of the customer" as opt-in consent from the customer. Subsequently, the Tenth Circuit vacated the rule's opt-in requirement because of First Amendment concerns. In October 2001, the FCC issued a notice seeking comment as to whether the FCC could meet the constitutional test and therefore should adopt opt-in consent, or should instead adopt opt-out consent for the carriers' use of CPNI under Section 222(c)(1). In addition, recognizing that two subsections of the Telecommunications Act regulate location information using different language ("express prior authorization" versus "approval of the customer"), the notice sought comment on what effect, if any, the provisions of section 222(f) have on the FCC's interpretation of section 222(c)(1). The FCC is currently reviewing the comments that it received and has not yet issued a final rule.

Even in the absence of final regulations interpreting these provisions, the statutory language itself clearly provides certain protections for location information and CPNI namely, that in most situations carriers obtain some form of consent before disclosing a consumer's phone records to a third party. One panelist noted that even these protections are limited, however, because they impose restrictions on common carriers but place no limitations on re-disclosure by third parties, such as wireless content providers, that receive the location information from common carriers.

Tracking Visits to Wireless Websites

Panelists stated that, as with surfing the wired Internet, users' browsing patterns on the wireless Web may be monitored and traced to individuals. As an individual surfs the mobile Web, the carrier and third-party content providers can collect a unique identifier, and possibly even the consumer's mobile phone number. Panelists pointed out that a person's wireless device tends to be strongly tied to an individual, even more so than a computer typically is, because people are less likely to share a wireless device with others; therefore, the potential consequences of an individual being tracked on the wireless Web are more significant than if the user were being tracked on the wired Internet.

A representative of a wireless content provider recognized that even companies that are not subject to FCC regulations (such as content providers, which are not common carriers under the statute) want to build trust with consumers, and they can do so by keeping consumers anonymous instead of tracking them on an identifiable basis. A representative of Sprint PCS stated that Sprint encrypts the user's telephone number when the user is surfing the wireless Web so that the user can remain anonymous. A consumer advocate urged companies not to use consumers' unique identifiers if they do not need to do so. Another panelist stated that numerous parties, including content providers and software developers, will need to work together to make sure that the user's phone number or a unique identification number will not be broadcast to every wireless website a consumer visits.

Greater Personal Data Collection

Panelists stated that wireless content is frequently most valuable to consumers when it is personalized. Therefore, businesses may seek to collect large amounts of highly personal information for use in personalization. Panelists predicted that wireless advertising, in particular, will be highly targeted and will focus on customer retention and customer relationship management instead of focusing on recruiting new customers. Panelists recognized, however, that collecting large amounts of highly personal information provides opportunities for abuse by the companies collecting it. In addition, law enforcement agencies may seek access to such information in criminal investigations, and civil litigants may seek access to the same information through civil process. Thus, some panelists suggested that companies should not collect information that they do not need, such as Social Security numbers.

Addressing Privacy Concerns in the Wireless Space

Participants provided numerous suggestions to address privacy concerns in the wireless space. This section will discuss recommended privacy protections and ways to build privacy into the technological architecture.

Privacy Protections for Wireless Services

Many workshop participants agreed that the intrusive and costly nature of offers and solicitations in the wireless space warrant special protections. A representative of the Center for Democracy and Technology stated that the information that wireless devices can collect is extremely sensitive. One panelist noted that it would be very problematic if a consumer gets unwanted solicitations on a wireless device or if the information gets into the wrong hands. Another panelist stated that consumers want privacy protections for information collected through wireless devices. Therefore, many workshop participants agreed that consumers should be given some form of control over information collected and used in the wireless space.

Panelists pointed out, however, that providing effective notice about information practices will be challenging, in a practical sense, because the screens on most wireless devices are so small that privacy policies are difficult to read. Many cellular phones only allow eighteen characters of text to be seen at a time, and PDA screens are relatively small as well. Thus, the tension between writing a comprehensive privacy policy and a policy that is concise and understandable will be exacerbated in the wireless space.

Panelists suggested that carriers, at least, can make privacy disclosures in the initial service contracts for wireless devices. Websites and other content providers, however, may or may not require consumers to sign service contracts before providing services. Panelists discussed some technologies that could help both carriers and content providers to provide effective notice. Some noted that "call-through" technology which allows a cellular phone user to click a link on a wireless website that automatically dials a phone number and connects the user to a live person or a recording may be useful in providing privacy disclosures. A representative of the privacy self-regulatory group, TRUSTe, also suggested that a combination of symbols, call-through technology, and disclosures in the wireless service contract could provide adequate privacy disclosures for wireless services. He cautioned, however, that a site that relied on audio disclosures would have to provide a mechanism for recording a written version of the stated policy on the date the consumer used the service, as well as provide a dispute resolution process, since a voice-based policy would not be a permanent mechanism.

A panelist from an Internet advertising company stated that privacy disclosures should be posted in several different places, and they should be clear, robust, and easy to understand. Another panelist stated that posting long privacy disclosures on wireless websites would likely be obtrusive. He therefore suggested that businesses consider posting short versions of their policies on their wireless websites, and making a more comprehensive disclosure available on a wired website or through an 800 number as long as these alternatives are made clear to the consumer using the wireless device. Panelists also recognized that many wireless transactions, such as viewing a bank account balance, will take place after the consumer registers for the service on a wired Internet site; in those instances, the consumer could first review the privacy policy on a larger screen on the wired Internet site.

Panelists also agreed that consumers should be given some choice about the collection and use of their information in the wireless space. Panelists stated that consumers will not like unsolicited messages on their wireless devices, although many panelists acknowledged that some companies will nevertheless send such messages. A representative from a wireless advertising technology company stated that advertisers should get opt-in consent before sending "push" advertising (where advertising content is sent to a wireless device at a time other than when the user requests it). However, he did not support opt-in consent for "pull" advertising (where the user goes to a website to receive information and receives an advertisement with the information), because it would be annoying to the consumer to have to agree to each advertisement that is served with requested content.

Regarding advertising and tracking of wireless Web users' surfing habits, one panelist suggested that users should be given two options when they first sign a contract for wireless service: 1) to pay for the content without any advertising, or 2) to get free content by agreeing to accept advertising and allow advertisers to track their website surfing and purchases, so that advertisers can profit from the advertising.

Many panelists supported added protections when location information is at issue. One consumer advocate stated that any consent to disclose location information must be truly informed consent, in order to give consumers confidence in the medium. He advised companies to set the default on wireless devices and systems so that there is no tracking of the consumer's location. Thus, consumers would have to activate the devices to enable location tracking (although the default should be overridden for 911 calls, so that people can be located in emergencies). A representative of ClickaDeal.com, a company that plans to provide location-based coupons, suggested that companies routinely purge users' location information. He stated that his company plans to cleanse its logs of user location information every hour so that it does not retain a history of a consumer's physical movements.

A representative of the Center for Democracy and Technology cautioned that ensuring meaningful choice in the wireless environment will be challenging for several reasons. First, the consumer may not know exactly what parties are potentially receiving personal information, because carriers, advertisers, and other service and content providers may be involved but not visible to consumers. Second, even opt-in consent, if it is provided only at the initial point in time when a consumer signs a service contract, and not at the point of information collection, may not be adequate; this is especially true if the privacy disclosures are not clear and easy to understand. The panelist stated rather than debating the concept of opt-in versus opt-out consent, companies should try to achieve informed consent that truly gives consumers control over their personal information.

Panelists agreed that manufacturers, carriers, content providers, and software developers will have to work together to enable effective notice and choice for consumers. Panelists also briefly discussed whether consumers should be given access to the information collected about them for review, correction, and/or deletion. Panelists stated that this is a complicated issue that raises difficult problems, such as how companies can authenticate users that request access to their information. Finally, panelists agreed that consumer education is an important component of protecting consumer privacy.

Building Privacy Solutions into the Technological Architecture

There was general consensus that technology can help improve privacy in the wireless space in numerous ways. Panelists suggested that implementation of the Platform for Privacy Preferences ("P3P"), a set of software-writing guidelines developed for the wired Internet by the World Wide Web Consortium, would be useful to consumers in the wireless space. P3P provides a language to express privacy policies in a machine-readable format. If P3P were implemented on wireless websites, a site would be able to express its information practices in P3P, and a P3P-enabled browser could read the P3P-enabled policy; thus, a user would not have to read a privacy policy on the device's small screen.

Beyond P3P, panelists also discussed a digital rights management approach, which would enable consumers to determine specifically what parties had access to their data, and provide technology so that consumers' permissions could be attached to their data. A representative of Nextel Communications stated that privacy could be enhanced by the use of a proxy or agent that acts on the user's behalf to enable previously set privacy preferences. A representative from Microsoft Corporation expressed support for "persona management," which would allow a user to provide different privacy preferences for Web browsing at different times. For example, a consumer could choose to be a "work persona" and provide only work contact information; a "home persona" and provide only home contact information; or an "anonymous persona" so that no personal information would be transmitted.

Security Concerns

Some panelists also expressed concern about the security of data transmitted through wireless devices. Panelists explained that although the public has the perception that wireless communications are vulnerable to interception over the airwaves, the risk of such eavesdropping is in fact very small. The more significant vulnerability exists within the carrier networks, especially at the point where the transmissions are translated from the wireless protocol (a set of rules governing wireless communications) to the wireline protocols that govern wireline communications. Other vulnerabilities exist once the transmission arrives at the wired Internet and becomes subject to the security vulnerabilities of the wired Internet.

Wireless devices are also easy to misplace and relatively easy to steal. As wireless devices become capable of storing more information, and conducting more sophisticated information processing, consumers have more information at stake if they lose their devices. Consumers already store highly sensitive information in their wireless devices; for example, many doctors store detailed patient information in their handheld devices. Thus, losing the device could compromise highly sensitive information and increase the risk of identity theft to the owner and others.

Panelists suggested that strong authentication procedures should be in place to prevent security breaches. They pointed out that such security systems should be transparent and intuitive to users. Currently, many cellular phones are enabled with locks to prevent unauthorized access to them, but most consumers do not routinely use the lock function because it is not convenient. Advances in technology could allow the owner to lock the device from a remote location if it is lost, even if the user did not lock the device beforehand. Security features also need to be affordable, and some security features may be so expensive that adding them would dissuade people from buying wireless products.

In addition, a representative of the technology infrastructure company MEconomy, Inc. made several recommendations to improve wireless security, including: 1) using an open platform for devices so users can load and unload their own privacy and security technologies; 2) separating personal identifiers from transactional data to increase privacy and security; and 3) only using data collected for a transaction for the specific transaction at hand. Other suggestions for improving wireless security were the implementation of authentication using public key infrastructure and the implementation of Wireless Transport Layer Security. In Europe, a standards working group is developing a small graphic that could be displayed on a phone to show that the transaction is secure, and one participant stated that this approach could be useful in the U.S. as well.

A panelist stated that if consumers purchase items over their mobile devices with a payment mechanism other than a credit card, such as billing a transaction directly to a consumer's wireless phone bill, the purchases are not protected by the Fair Credit Billing Act; thus, there would be no legal requirement that the seller provide a mechanism for dispute resolution if the consumer alleges that unauthorized charges were made through the device. The panelist recommended that if possible, consumers should use their credit cards to engage in m-commerce transactions.

Chapter 8

Experts Say Cell Sites Are Safe; Some Security in Place

A few weeks ago, a technical worker tried to measure radio-frequency emissions from a cell site mounted on a water tower walkway. Before he was able to climb the tower, however, a police cruiser pulled up and officers asked him what he was doing.

"We get those kinds of situations," says Wes McGee, president of SiteSafe Inc., a Crown Castle International Corp. subsidiary that makes sure cell sites on buildings and elsewhere meet health and safety requirements.

McGee contends such incidents are rare and cell sites and telecom towers are and always have been secure. But episodes like this may occur with more frequency in the future. In years past, carriers, tower companies and cell site managers mostly worried about natural disasters or the occasional vandal. Now, however, the specter of more terrorist attacks has added an additional layer of concern in some people's view.

So far, increased concern has not led to action. Despite warnings from some industry experts that the nation's communications networks need increased protection, McGee and others say no major security procedures have been put in place since the September 11 terrorist attacks. A few minor changes have been made, however. New York law enforcement officials, for instance, now demand to see a company photo I.D. from anyone working around the sites for any reason.

"Experts: Cell Sites Safe; Some Security In Place," by Deborah Méndez-Wilson, Wireless Week.com, http://www.wirelessweek.com/index.asp?layout=print_page&articleID=&doc_id=56488, November 14, 2001. Reprinted with permission of Wireless Week.

"For the most part, we do not see any increase in security. If you pick specific sites, you can, from time to time, find someone who has implemented a sign-in procedure," McGee says. "Apparently, [security] has not been as big an issue as most people think."

"Most tower sites are gated and have access limitations anyway. Rooftops, for the most part, are locked. It really depends on the site," McGee adds.

But not everyone agrees the sites are secure enough. Shortly after the attacks, a few industry executives—including Larry Babbio, vice chairman of Verizon Communications, and Joseph Nacchio, chairman and CEO of Qwest Communications International Inc.—warned federal lawmakers and regulators that the nation's telecom networks could be vulnerable to attack.

A Verizon switching site near the World Trade Center endured heavy damage when the towers collapsed. The damage—as well as the loss of equipment on one of the towers—disrupted landline and wireless phone service in Manhattan for days and was blamed for contributing to the shutdown of the New York Stock Exchange for an unprecedented four days. During a tour of the damaged switching site, Babbio reportedly said communications would come to a halt in the United States if terrorists were to disrupt the operations of the 50 largest switching centers in the country.

"If you really want to create a panic in this country, you take down the telecom network," Babbio said, according to one published report.

For his part, Nacchio testified before the U.S. Senate Government Affairs Committee on Oct. 4, telling lawmakers the U.S. telecom networks are strong, but that the Bush administration and Congress "should take additional steps to protect the security of all our country's critical public and private network facilities."

The committee's hearing focused on protecting critical telecom and other infrastructure from cyber and physical attacks. Nacchio is vice chairman of the National Security Telecommunications Advisory Committee, which provides the president with advice on security and emergency preparedness issues.

On September 11, when Verizon's infrastructure was damaged, thousands of wireless subscribers saw how easily phone service could be disrupted. The switching center that was severely damaged served more than 200,000 access lines and more than 3 million private line circuits. However, increased demand for phone and Internet service following the attacks caused the lion's share of difficulties, according to the Yankee Group, a high-tech market research firm based in Boston.

Call volumes over landline and wireless networks rose dramatically, in some cases as much as 200 percent above normal. Web sites were frozen as users flooded the Net. All in all, however, "the wireless network showed remarkable strength and survivability," Yankee Group analysts concluded in a special report. Analysts commended Verizon Wireless and other carriers for deploying cells on wheels, or COWs, to improve coverage and capacity in lower Manhattan as soon as safety precautions permitted. Some of Crown Castle's antennas affixed to a tower located on the other side of the Hudson River also were reconfigured to enable Verizon Wireless to increase its network capacity.

Few carriers, however, are willing to discuss any new security measures they might be putting in place throughout their networks as a result of the terrorist attacks. "We take the security and safety of our assets—be that our people, facilities or equipment—very seriously. For precisely that reason, we don't talk about or comment on our security procedures," says Alexa Graf, an AT&T Wireless spokeswoman.

SiteSafe's McGee says utility companies have started to "harden up on telecom sites" to prevent possible infiltration by saboteurs. SiteSafe performs a lot of work for the United Telecom Council, which represents many U.S. utilities. Most utility companies rely on private two-way radio communication networks. Utilities are looking at access issues, particularly at power plants, McGee says.

Crown Castle, which owns and manages more than 15,000 tower sites internationally, has been asked to notify natural gas companies before engineers are sent to cell sites located near pipelines, says David Binstock, Crown Castle's director of operations. If either the tower managers or a wireless service provider needs to visit the site for whatever reason, the utility companies now request them to call ahead and let them know who's coming and when, he says.

Binstock says security for the most part has been beefed up around power plants, water reservoirs, high-rise buildings and other potentially sensitive sites. Because of that, Crown Castle has asked field workers to obtain company photo badges and show them when necessary.

Some tower and cell site management companies use remote locking solutions to ensure that unauthorized people are unable to get into restricted areas. Others have systems in place that require employees to scan electronic I.D. cards or enter passcodes. Crown Castle is looking at several new solutions that would help increase the security of the company's physical assets, wherever they might be, Binstock says.

In any case, damage to a single tower site isn't likely to cause widespread disruption of wireless service. When people talk about network security, "they are talking about their major switch centers, where lots of voice and data information" flows, Binstock says.

Observers who point to the vulnerability of such telecom hubs say service providers, corporations and workers must adopt new ways to communicate in light of the terrorist attacks. Some argue that telecom networks should be more dispersed and that companies should explore modes of communication that won't tie employees to landline or wireless phones, including messaging and video-conferencing solutions.

Frank Boscarillo, senior vice president of network operations for the Bedminster, N.J.-based KMC Telecom, a nationwide provider of fiber-based integrated communications, recommends companies be prepared for what may come. "By designing a plan that is meant to react to the worst-case scenario, you will ensure measures are incorporated [that] can be applied to lesser emergencies," he says.

As the Yankee Group concludes: "There is security in diversity." In the case of wireless networks, that philosophy might prove true now more than ever.

Chapter 9

Smartphones: Consumer Convenience and Emergency Assistance or Privacy Invasion?

One of the most often-mentioned reasons for acquiring a wireless telephone is for personal safety. The Federal Communications Commission's "E911" rules require wireless telephone companies to precisely locate 911 callers as an aid to finding them in emergencies. These carriers must upgrade their networks in order to ensure that they deliver, with stipulated accuracy, the callers' geographical coordinates to public safety dispatchers. (Carriers are upgrading now, but some dispute the FCC's timetable for implementing user location capability.)

This vast, mandatory investment will make no money for wireless carriers if it is limited to emergency use. Calls to 911 are free of charge. To generate revenue from this capability, the industry is deeply engaged in developing commercial applications that depend on locating citizen users. Some phones will place a "locate me now" button under the control of the user, who may choose where and when to be located.

Reprinted from "Consumer Convenience and Emergency Assistance or Privacy Invasion?" by Bennett Kobb, *The Forum Connection*, September 2001. The Civil Rights Forum on Communications Policy works to bring civil rights organizations and community groups into the debate over the future of our media environment—an environment that is the key to the future of the nation. *The Forum Connection* is a publication that seeks to update readers on important developments in communications policy with a focus on equity and civic participation. This service is made available through the support of the Ford Foundation, the W.K. Kellogg Foundation, the Albert A. List Foundation, and the John D. And Catherine T. MacArthur Foundation. Document location: http://www.civilrightsforum.org/tech092001.html.

But this will require purchase of a new phone. A competing technology is "network-based" location; it works with current phones. Subscribers of network-based wireless phone services can always be located, as long as the phone is on.

"There are numerous areas in which location data can be utilized," according to an industry group, the Location Interoperability Forum. "For example, a user can review the menus of nearby restaurants over [the phone screen], receive advertising according to personalized profiles, or find out if there are any friends around in town."

Wireless locating technologies "enable the creation of detailed daily itineraries for millions of consumers," according to the Electronic Privacy Information Center.

A Petition to Protect Citizens

In November 2000, the Cellular Telecommunications and Internet Association (CTIA), which represents wireless carriers, petitioned the FCC to require that citizens receive prior notice of "location information collection and use practices" before the carriers collect that information; have a meaningful opportunity to consent to collection and use of location-based information; and be assured of the "security and integrity" of such information. At press time the FCC had taken no action on the CTIA petition other than to accept public comments.

CTIA asked the FCC to adopt the association's privacy "principles," which require that a wireless carrier provide notice (via e-mail, websites or billing inserts) about its location-collecting practices. The carrier would have to enable customers to consent to being located, and to the carrier's use or disclosure of the location records. Carriers and their third-party business partners would have to maintain the location information "securely," protected from unauthorized release or use.

Civil rights advocates should be concerned about the effectiveness of these mechanisms—and the implications of a network that can track and exploit the whereabouts of citizens on a moment-to-moment basis.

Implications for Civil Rights

- Will law enforcement authorities have access to location data? With E911, public safety agencies will automatically receive highly accurate descriptors of caller location—in times of emergency. Could police track citizens whenever desired, not just in emergencies? Could judges obtain the travel records of persons under restraining order?

- Who has access to customer location data? Only FCC-licensed wireless carriers, or third parties such as "concierge" services, towing companies and market researchers? Will private individuals be able to purchase the instant whereabouts of any person of interest? Should government control these parties' use of location information?
- A database could reveal when and where two or more users of wireless devices were in physical proximity. How will this proximity information be recorded and used? Do you want others to know who you are meeting with?
- Will those who withhold consent to tracking be charged higher phone rates?
- What is a meaningful opportunity to consent? Will it be "opt-in" or "opt-out," requiring specific action such as answering a letter or visiting a website?
- Does using a phone to request a service such as road directions imply consent to being located?
- How long does the consent last—only long enough to conduct a location-related transaction, or months or years? How long is the location information stored? Can it be resold?
- What if a user forgets about past consents? Will "privacy providers" manage users' location permissions and relationships for a fee?
- Will access to insurance reflect where one travels or which other wireless users one associates with? (Rental car companies are tracking customer travel and speed, and charging extra for exceeding specified limits. Some auto insurers offer lower rates for drivers who consent to electronic tracking and stay out of high-risk areas. Could similar practices apply to pocket phone users?)
- Will users have the right to inspect the location data others have gathered about them, and to correct it if necessary? Will they have to pay for such access? Will the presented data be comprehensible?

Just A Friendly Tracking Device

SiRF Technology, a supplier of location devices, said that "A location-enabled phone could identify the specific location from which an individual sought information. For example, it could identify the apartment that the user was in when activating a service to find out about nearby entertainment in the middle of the night. Unless the individual can show that the phone was loaned or stolen, the phone owner is almost irrefutably placed at the scene.

"Over time, the surprise and unfairness of being confronted with embarrassing or incriminating information in such detail increases," the company said. "A single incident in which a lawyer used five-year old service records to contradict the sincere testimony of a court witness could trigger a consumer backlash so severe that it could even threaten consumer acceptance of all location services, even E911."

"The hollow, non-binding principles proposed by CTIA offer no protection to the consumer," stated the Wireless Consumers Alliance. "Notice is not notice when buried in other documents or letters or when a consumer must take steps to learn of the invasion of their privacy. Consent is not consent when hidden in 'agreements' or on web pages."

The Alliance urged the FCC to ensure that providers retain customer location data no longer than necessary to provide a requested service; that "record-keeping practices shall be open and no record systems shall be secret," and that users have access to records kept concerning their activities. The FCC should not pre-empt state privacy laws, it argued, and "Actual and consequential damages should be available to the consumer for violation of the rules."

The Electronic Privacy Information Center (EPIC) had a similar view: "The FCC should ensure that consumers consent prior to the collection of information that will be stored because that information is obtainable by government search warrants and subpoenas. "The likelihood that someone's electronic records might later be accessed pursuant to a warrant or subpoena are not remote," it said, citing a study that found an 800% increase in law enforcement demands for subscriber information from Internet providers.

Who Can the Government Protect You from?

A key issue in the debate is whether the FCC could regulate privacy practices of wireless location information providers that are not Commission-licensed carriers.

Sprint Corp. said that the FCC "must recognize that it does not, today, have the tools to address the subject [of location privacy] comprehensively" in part because any rules it adopts "would extend only to wireless carriers and not to the hundreds of entities that will have access to location information."

For example, the American Automobile Association operates a road-assistance service called "Response". Subscribers to this service can press a button on their wireless phones, which transmits the user's satellite-derived location to AAA. The association believes that the FCC can regulate carriers but not "end-users" who may obtain and use customer information, though, it said, the practices of end-users could be subject to review by the Federal Trade Commission or the states.

It seems likely that services other than conventional wireless phone companies will be part of the location-finding apparatus—not only so-called "end users" like AAA, but other technologies will play a role.

Bluetooth

For example, "Bluetooth," a short-range wireless "personal area network," is expected to be included in millions of wireless phones, computers, and other consumer devices over the next several years. Bluetooth devices used in the U.S. must be FCC-authorized, but no licenses are involved.

According to the Center for Democracy and Technology (CDT), "Bluetooth-enabled devices will support services such as money transfers between individuals and networking. Each Bluetooth-enabled interaction will indirectly reveal the individual's physical location to the device, and corresponding individual or entity, with which it interacts.

"The location information created by these various interactions can be collected and used by a variety of companies and later accessed by private parties or government agencies without the individual's knowledge or consent."

Ericsson Corp. has recently started marketing the $499 "Blip," (Bluetooth Local Infotainment Point) a sophisticated wireless server inside a small unit to be installed wherever people gather. Blip will provide handheld device users with web pages and information of local interest, and could conceivably record the identities and locations of users if outfitted with the necessary software.

Bluetooth was originally developed for "cable replacement," substituting radio waves for the strings of wires that connect computers and their peripheral devices such as printers and office servers. But developers are banking on Bluetooth outside the business environment. A company called Tripda, for example, offers Bluetooth-based "private and controllable wireless data and voice communities" for stadiums, theaters, amusement parks, restaurants, malls, stores, zoos, and museums, in addition to factories and offices.

A Proposal from Senator Edwards

In what appears to be an effort to resolve regulatory ambiguity, Sen. John Edwards (D-NC) introduced the Location Privacy Protection Act (S-1164) on July 11, 2001. (There was no companion bill in the House at presstime.)

"There is a substantial Federal interest in safeguarding the privacy right of customers of location-based services or applications to control the collection, use, retention of, disclosure of, and access to their location information," the bill states. "Location information is nonpublic information that can be misused to commit fraud, to harass consumers with unwanted messages, to draw embarrassing or inaccurate inferences about them, or to discriminate against them. Improper disclosure of or access to location information could also place a person in physical danger. For example, location information could be misused by stalkers or by domestic abusers."

The bill closely tracks CTIA's privacy principles. It would require the FCC to adopt rules to ensure that providers of location-based services and applications inform customers, with "clear and conspicuous notice," about their policies on the collection, use, disclosure of, retention of, and access to customer location information.

Except for such purposes as billing, network protection and providing the user's requested service, providers of location-based services and applications would have to obtain a customer's express authorization before "collecting, using, or retaining the customer's location information" or "disclosing or permitting access to the customer's location information to any person who is not a party to, or who is not necessary to the performance of, the service contract between the customer and such provider." The bill would pre-empt any local or state regulation that conflicts with these provisions.

Importantly, the bill would apply "to any person that provides a location-based service or application, whether or not such person is also a provider of commercial mobile service" (namely, an FCC-licensed mobile communications carrier such as a cellular telephone company).

In this respect, the proposed Location Privacy Protection Act would reach not only the telecommunications carriers, but other, myriad players in the location chain not historically under FCC jurisdiction. Depending on interpretation, the term "application" could include operators of location-based services or keepers of databases—companies that own no wireless facilities and hold no applicable FCC licenses.

John Jimison, executive director of the Wireless Location Industry Association (WLIA), told us that "The FCC can and should provide standards for the regulated carriers, as CTIA has requested, but I doubt if WLIA will agree that the FCC should have its authority extended to cover software companies, hardware manufacturers, advertisers, asset-tracking services, automobile companies, fleet managers, or others in the diverse group of companies who are developing commercial applications from the new technology for wireless location services. As long as they understand and do not violate basic privacy rights of users, they should be free to develop their businesses."

The civil rights community should welcome efforts to obtain FCC oversight of the location information industry, but the diverse implications for privacy invasion—and the number and types of players who may enter this field—could exceed that agency's resources and even its charter, forcing a rollback of this legislation's scope.

Government and private surveillance has a direct impact on a wide variety of traditional civil rights concerns. If there is a right to travel and to privacy of association, is there a right to keep one's travel and associations private? The political temperature of these questions will increase with bulging subscriber rolls. More than 120 million U.S. citizens subscribe to wireless phone service, and thousands more start service each day.

Chapter 10

As Cell Phones Get Smarter, They Become Prone to Viruses

For malicious computer hackers and virus writers, the next frontier in mischief is the mobile phone.

A phone virus or "Trojan horse" program might instruct your phone to do extraordinary things, computer security experts say. "If a malicious piece of code gets control of your phone, it can do everything you can do," said Ari Hypponen, chief technical officer of Helsinki-based F-Secure, a computer security firm. "It can call toll numbers. It can get your messages and send them elsewhere. It can record your passwords."

As cellular phones morph into computer-like "smartphones" able to surf the Web, send e-mail and download software, they're prone to the same tribulations that have waylaid computers over the past decade. "We should think of cell phones as just another set of computers on the Internet," said Stephen Trilling, director of research at antivirus software maker Symantec. "If they're connected to the Internet they can be used to transmit threats and attack targets, just as any computer can. It's technically possible right now."

In Japan, deviant e-mail messages sent to cell phones contained an Internet link that, when clicked, caused phones to repeatedly dial the national emergency number—equivalent to 9-1-1. The wireless carrier halted all emergency calls until the bug was removed.

"Hackers' next target? Cell phones: as Mobile Devices Get 'Smarter,' They Become Prone to Viruses," by Jim Krane, Associated Press, March 10, 2002. Used with permission.

In Europe, handsets' short message service, or SMS, has been used to randomly send pieces of binary code that crashes phones, forcing the user to detach the battery and reboot. A new, more sinister version keeps crashing the phone until the SMS message is deleted from the carrier's server.

In the United States, relatively primitive cell phone technology keeps users immune from such tricks, for now.

Phone hacking is nothing new. In the 1970s, so-called "phone phreakers" made free phone calls—and even gained control of major phone trunk lines—by whistling certain tones into the receiver. "It was easy," said John Draper, 58, of Stockton. Draper, now a designer of computer security software, is still known as Captain Crunch for pioneering the hacking of phone networks with the help of a plastic whistle that came in a box of the eponymous breakfast cereal. "You could control the entire network, do anything an operator could do," Draper said.

Now, at least three software companies have released personal security software for emerging smartphones, girding for a new wave of phone viruses and Captain Crunch-style tricks. Hypponen's F-Secure is one such firm, selling anti-virus and encryption software for smartphone operating systems made by Palm, Microsoft and the Symbian platform common in Europe.

Thus far, there have been no publicized reports of phone hacking or viruses, although viruses have attacked handhelds running the Palm operating system. Microsoft predicts deviant code will soon emerge for handhelds running its Pocket PC software. Both operating systems are expected to be used increasingly in smartphones.

A virus is a piece of malevolent code that self-replicates, while a Trojan horse does not but can be just as destructive. The pranks in Europe and Japan created virus-like havoc, but did not propagate like a full-fledged virus. For virus writers who crave notoriety by wreaking maximum havoc, there are still too few smartphones, and no widespread software platform to attack, Hypponen said. That is starting to change.

Until recently, cell phone operating systems were "closed," unable to download software. But new smartphones—like the Nokia Communicator, Handspring's Treo, Motorola's Java Phone and Mitsubishi's Trium-Mondo—are open to such third-party downloads. At the same time, software developers' tools available for designers of such programs as games and currency converters can also be used to create malicious applications, Hypponen said. "It's possible for anyone to make custom software for this platform," he said. "Teens can download development tools and write their own software."

It's these third-party programs that worry experts. If one is disguised as a Trojan horse, an infected phone could make some calls on its own. In a speech at a cell phone conference in France last month, Hypponen cited a Slovak Web site, virus.cyberspace.sk, that posted a bulletin exhorting readers to create phone viruses. " 'We are starting Cell Phone Virus Challenge. Any contribution welcomed,' " Hypponen quoted the notice as saying. The page has since been taken down.

Soon, mobile phone owners will be obliged to install security software like "personal firewalls" that used to be reserved for Internet servers, said Prakash Panjwani, a senior vice president at Certicom, a computer security firm in Hayward. "That's where things are going," said Panjwani. "It's the same threat as the wired world: people posing as you, stealing your identity or your personal information, and using your information for malicious purposes."

Cell phone users can avoid this, of course, by sticking with their old "dumb" phones, said Alan Reiter, a wireless consultant in Chevy Chase, Md. "There are trade-offs," said Reiter. "Do you want a phone with a tiny monochrome screen where you can only make phone calls? That's much more secure."

Chapter 11

Mobile Device Viruses: Nothing to Worry about?

Most sensible people watch out for the everyday threats they are most likely to encounter, rather than fear predicted occurrences that may never surface.

Do Cell Phone Risks Exist?

There has been much discussion recently about the potential vulnerabilities of new technologies, such as WAP (wireless application protocol) cell phones and palmtop computers. Concern has focused on whether or not these devices can be infected by viruses. Judging by the volume of press releases coming from some anti-virus vendors on these subjects, you would imagine mobile devices are at great threat of attack.

The fact is, to date, there is no virus that infects cell phones, despite the hysterical press releases, media stories and hoaxes stating the contrary. What have been seen are viruses that are capable of sending text (SMS) messages to cell phones. For instance, VBS/Timo-A is an e-mail-aware worm that can send text messages to phones. Another infamous virus, the LoveBug, is capable of forwarding its code to fax machines and cell phones via Microsoft Outlook. Of course, neither of these viruses causes any harm to the mobile devices and both are incapable of spreading further.

Excerpted from "Mobile Device Viruses: Nothing to Worry About?," SC Online, SC Magazine, 161 Worcester Road, Suite 201, Framingham, MA 01701, http://www.scmagazine.com/scmagazine/sc-online/2000/mobile_device_viruses/article.html. December 2000.

A growth area for mobile communications is in wireless application protocol (WAP). WAP is based on the same model as web communications in that a central server delivers code, which is run by a browser installed on the phone or organizer. It's important to note, though, that there is nowhere on current WAP devices where a virus can harbor itself. Unlike a PC, a WAP phone is not able to store the applications it uses.

Also, there is no way a virus would be able to spread to other WAP users. Current WAP-enabled cell phones do not allow for communication between "client" phones. Simply put, code passes from the phone company's server down to your phone, but not vice versa or from one phone to another.

The bottom line in this case is that cell phones and WAP-enabled devices are simply not sophisticated enough to be infected at the present time. However, consumer demand for increased functionality often means that manufacturers are keen to develop the technology required to meet user requirements. As these devices become more complex, the opportunities for viruses to infect them may also rise.

PDAs Pose Problems?

What about palmtop computers and personal digital assistants (PDAs)—can they be infected by computer viruses?

PDAs run specially written scaled-down operating systems, such as EPOC, PalmOS and PocketPC. They are often connected to home or office PCs to synchronize the data on the two machines. This presents an opportunity for viruses to spread onto them.

Yet, no viruses currently exist for the PocketPC and EPOC operating systems, although there is no technical reason why they could not be written. There is a virus called Palm/Phage which is able to infect Palm OS, but it is not in the wild and poses little threat. Nonetheless, it is sensible to keep backups of any Palm applications and data.

There is also a Trojan horse known as Palm/Liberty-A, which is able to infect the Palm OS. It deletes Palm OS applications and was distributed in the "warez" community. Like Phage, it is low risk and you are unlikely to ever encounter it.

Viruses Make a Bid for Bluetooth?

Bluetooth is a standard for low-power radio data communication over very short distances. Computers, mobiles, fax machines and even domestic appliances, like video recorders, can use Bluetooth to discover what services are provided by other nearby devices and to establish transparent links with them.

Software that utilizes Bluetooth is currently emerging. Sun's Jini technology allows devices to form connections, exchange Java code automatically and give remote control of services. The worry is that an unauthorized user, or malicious code, could exploit Bluetooth to interfere with these services.

However, Bluetooth and Jini are designed to ensure that only trusted code from known sources can carry out sensitive operations. This means that it is highly unlikely for a virus outbreak to occur.

What's to Happen?

Inevitably, the evolution of mobile and PDA technology will bring with it the development of further security. The issue here is where you implement anti-virus measures.

The most efficient way to protect mobile devices is to check data when you transfer it to or from the device. For cell phones, the WAP gateway would be a good place to install virus protection. All communications pass through this gateway, providing an ideal opportunity for virus scanning.

As cell phones become increasingly interconnected, it will be difficult to police data transfer at a central point. Then, the solution will be to put anti-virus software on individual phones—that is, once they have sufficient processing power and memory.

In the case of PDAs, anti-virus software could be used during data synchronization with a conventional PC, but again, there will be an increasing requirement for anti-virus on the PDA itself.

It is easy to get carried away with the potential virus threat. However, much of the hype is unsubstantiated and based on mere speculation. There have been some ludicrous suggestions about viruses. At the moment mobile devices are just not sophisticated enough to allow widespread virus infection. A virus is limited by the functionality of the platform it infects.

The current trend seems to be for people to worry about the potential threats of tomorrow, which may never come to fruition, as opposed to the real risks of today. The best advice to follow is to remain alert to what the dangers are right here, right now and to protect against them. While you're concerning yourself about the future, you could be missing what's right under your nose.

—by Graham Cluley

Graham Cluley is head of corporate communications for Sophos Anti-Virus (www.sophos.com).

Chapter 12

Global Positioning System at Risk

A study released the week of January 11, 2002 urges the Bush administration to take immediate steps to prevent terrorists from crippling the satellite-based Global Positioning System (GPS), a key technology in battlefield systems, navigation systems and other critical applications across government.

The report, "Defending the American Homeland," from the Heritage Foundation, a Washington, D.C.-based think tank, recommends President Bush designate GPS radio frequencies and network systems as critical national infrastructure, so these systems receive the same protection given to telecommunications, financial systems, utilities and other core operations of vital interest to the country.

The report also recommends assigning responsibility for its security to the Department of Defense and taking immediate steps to make the network more secure.

The report's other recommendations include:

- Secure all federal networks and information systems, which would require revising agencies' technology-purchasing guidelines to place a premium on security, as well as exploring alternatives to the proposed GovNet system, which would move

---

"Report: GPS at risk," by Dan Caterinicchia, Federal Computer Week, http://www.fcw.com/fcw/articles/2002/0107/web-heritage-01-11-02.asp, January 11, 2002. Reproduced with permission of *FCW.COM*,

critical government systems off the public Internet and onto a private Internet-like network.

- Rapidly improve information-gathering capabilities at all government levels, which should include having the Office of Homeland Security establish a group to develop a national strategy for gathering and sharing intelligence.
- Improve intelligence and information sharing among all government levels through the creation of a federal information fusion center where all intelligence data is sent and from where it is dispensed on a need-to-know basis.
- Develop a program to increase airport and seaport controls, including a federal interagency center to analyze the people and products entering the U.S. by sea.
- Direct the military to aid federal, state and local officials in their counterterrorism efforts by identifying critical infrastructure nodes, assessing security levels, and providing protection for them, as well as the redundant communications, command and control systems.

Michael Scardaville, policy analyst for homeland security at the Heritage Foundation and a member of the task force, said its plan is to promote its findings to the administration, Congress and state and local governments to "get as many recommendations instituted as possible."

Scardaville said the driving factors of all the report's recommendations were that they could be done "relatively quickly, at a reasonable cost, and without any major shake-ups." He said that adding GPS to the nation's critical infrastructure is a perfect example because it simply requires a presidential directive mandating it "and that is not a difficult thing to do."

The Heritage Foundation sent copies of the report to the White House and some members of Congress, but has yet to receive official feedback, Scardaville said.

Chapter 13

Wireless Payment Technology Uses and Risks

We are entering a new era of payment technologies that could have you reaching for your mobile phone instead of your wallet at the check-out counter. The Mobile electronic Wallet (MeW) application will enable your mobile phone, PDA, Smartphone (combination PDA and mobile phone) to store and use financial tools (i.e. electronic credit cards, cash, checks, and receipts) at the point of sale, or Local Transactions (LTs). To initiate an electronic payment you simply press a button on your mobile device. Within seconds your transaction is completed and an electronic receipt is stored within your device.

There are several industry groups currently working to create a single set of standards for worldwide compatibility of mobile local transaction solutions. These groups include:

Infrared Data Association's (IrDA) Infrared Financial Messaging Special Interest Group (IrFM SIG)

Infrared Financial Messaging (IrFM) is an international standardization initiative of IrDA whose key participants are Palm, Nokia, Agilent, Ericsson, Motorola, Sharp, Infineon, Extended Systems, Hewlett-Packard, Zilog, Visa, Vishay Telefunken, Harex InfoTech, VeriFone, and CrossCheck. The IrFM protocol defines payment usage models, profiles, architecture, and protocol layers to enable hardware,

---

"Point & Shoot: The Future of Wireless Payment Technologies," by Jourdan Zayles, WebProNews, http://www.webpronews.com/2002/0314.html, March 14, 2002. Used with permission of iEntry, Inc., www.webpronews.com.

software, and systems designers to develop IrFM-compliant products that are globally compatible. Electronic wallet applications where users beam their financial information to make a payment and receive a digital receipt have been demonstrated using IR-enabled Palm and Handspring Visor PDAs.

Mobile Electronic Transaction Forum (MeT Forum)

MeT is an initiative started by Ericsson, Motorola, and Nokia to establish a framework for secure mobile transactions (ability to buy goods and services using a mobile device).

Bluetooth Special Interest Group's (Bluetooth SIG) Short Range Financial Transaction Study Group (SRFT SG)

The Bluetooth SIG SRFT Study Group is a research group within Bluetooth examining the development of a profile that improves the device discovery and connection times for short-range financial transactions at the point of sale. Key participants include Toshiba, Tech, Verifone, Panasonic, and IBM.

National Retail Federation (NRF)

The Association for Retail Technology Standards (ARTS) of the National Retail Federation is a retailer-driven membership organization dedicated to creating an international, barrier-free environment for retailers. ARTS was established in 1993 to ensure that technology works to enhance a retailer's ability to develop business solutions while maintaining industry standards that provide greater value at lower costs.

There has been much discussion among analysts about the use of RF (radio frequency), rather than IR (infrared), for short-range data transfer. One such wireless protocol that has been getting a lot of attention in the past few years is Bluetooth. Originally developed by Ericsson in 1994 and named for Harold Bluetooth, the Viking king who united Denmark and Norway in the 10th century, Bluetooth uses a short-wave, always-on radio signal that lets devices of all kinds communicate with one another, including mobile phones, printers, laptops, and PDAs. Since it uses RF waves, communication doesn't require a line-of-sight connection between devices, as does IR. Unlike IR, Bluetooth can travel through non-metal obstructions like furniture and walls, send signals in all directions, and be initiated by the devices themselves. The biggest buzz about Bluetooth is centered on the era of "personal area networks" that this technology promises to create. The thinking behind this is that once Bluetooth components become inexpensive they will be embedded in all kinds of machines, including washer-dryers, stoves, VCRs, and CD-players, all of which could be monitored and controlled by Bluetooth.

With the development of Infrared Financial Messaging (IrFM), a new "point and pay" wireless payment standard designed to accommodate 200-plus millions of IrDA-enabled devices already in the marketplace, it's becoming apparent that IR technology will be playing a major role in the future of financial transactions. The numerous advantages IR technology has over RF for local transactions include:

- **Security:** Because RF solutions (including Bluetooth and devices with wireless Ethernet connectivity) send radio waves in all directions, not point-to-point, they lack the inherent security offered by IR when implemented in transactions requiring confidentiality.
- **Ease of Use:** Another problem with RF solutions is that competing 802.11 standards haven't yet been formalized, whereas IR standards have been developed for a multitude of applications.
- **Market Presence:** Much of the necessary hardware for short-range RF technologies doesn't exist yet or hasn't penetrated the market in substantial quantity. IR hardware and software is readily available.
- **Speed:** The data transfer rate for IR solutions is five times higher than that of Bluetooth.
- **Power:** IR uses five times less power than Bluetooth.
- **Cost:** IR costs 10 times less than Bluetooth.

The future of wireless payment technologies will mean always having the right change for parking meters and vending machines, a quick and easy checkout experience, and easy access to receipts, coupons, and gift certificates. Paying for almost anything by pressing a button on your mobile phone or PDA will soon become a way of life. Because IR allows all types of devices, both costly and cheap, to share data, IR solutions suit both low-cost, high volume applications as well as high-end applications. Up until now the IrDA and Bluetooth payment solutions have been mere technology demos rather than mass market rollouts. Analysts predict that by the first quarter of 2002, local transaction technology should start rolling out to consumers.

—by Jourdan Zayles

Jourdan Zayles is a contributing writer at MedioCom.net.

Chapter 14

Is Wireless Phone Radiation Harmful?

Do Wireless Phones Pose a Health Hazard?

The available scientific evidence does not show that any health problems are associated with using wireless phones. There is no proof, however, that wireless phones are absolutely safe. Wireless phones emit low levels of radiofrequency energy (RF) in the microwave range while being used. They also emit very low levels of RF when in the stand-by mode. Whereas high levels of RF can produce health effects (by heating tissue), exposure to low level RF that does not produce heating effects causes no known adverse health effects. Many studies of low level RF exposures have not found any biological effects. Some studies have suggested that some biological effects may occur, but such findings have not been confirmed by additional research. In some cases, other researchers have had difficulty in reproducing those studies, or in determining the reasons for inconsistent results.

What Is Radiofrequency Energy (RF)?

Radiofrequency (RF) energy is another name for radio waves. It is one form of electromagnetic energy that makes up the electromagnetic spectrum. Some of the other forms of energy in the electromagnetic spectrum are gamma rays, x-rays and light. Electromagnetic energy

---

Excerpted from "Cell Phone Facts, Consumer Information on Wireless Phones, Questions and Answers," U.S. Food and Drug Administration (FDA), http://www.fda.gov/cellphones/qa.html. Updated April 3, 2002.

(or electromagnetic radiation) consists of waves of electric and magnetic energy moving together (radiating) through space. The area where these waves are found is called an electromagnetic field.

Radio waves are created due to the movement of electrical charges in antennas. As they are created, these waves radiate away from the antenna. All electromagnetic waves travel at the speed of light. The major differences between the different types of waves are the distances covered by one cycle of the wave and the number of waves that pass a certain point during a set time period. The wavelength is the distance covered by one cycle of a wave. The frequency is the number of waves passing a given point in one second. For any electromagnetic wave, the wavelength multiplied by the frequency equals the speed of light. The frequency of an RF signal is usually expressed in units called hertz (Hz). One Hz equals one wave per second. One kilohertz (kHz) equals one thousand waves per second, one megahertz (MHz) equals one million waves per second, and one gigahertz (GHz) equals one billion waves per second.

RF energy includes waves with frequencies ranging from about 3000 waves per second (3 kHz) to 300 billion waves per second (300 GHz). Microwaves are a subset of radio waves that have frequencies ranging from around 300 million waves per second (300 MHz) to three billion waves per second (3 GHz).

What Is FDA's Role Concerning the Safety of Wireless Phones?

Under the law, FDA does not review the safety of radiation-emitting consumer products such as wireless phones before they can be sold, as it does with new drugs or medical devices. However, the agency has authority to take action if wireless phones are shown to emit radiofrequency energy (RF) at a level that is hazardous to the user. In such a case, FDA could require the manufacturers of wireless phones to notify users of the health hazard and to repair, replace or recall the phones so that the hazard no longer exists.

Although the existing scientific data do not justify FDA regulatory actions, FDA has urged the wireless phone industry to take a number of steps, including the following:

- Support needed research into possible biological effects of RF of the type emitted by wireless phones;
- Design wireless phones in a way that minimizes any RF exposure to the user that is not necessary for device function; and

- Cooperate in providing users of wireless phones with the best possible information on possible effects of wireless phone use on human health.

FDA belongs to an interagency working group of the federal agencies that have responsibility for different aspects of RF safety to ensure coordinated efforts at the federal level. The following agencies belong to this working group:

- National Institute for Occupational Safety and Health (NIOSH)
- Environmental Protection Agency (EPA)
- Federal Communications Commission (FCC)
- Occupational Safety and Health Administration (OSHA)
- National Telecommunications and Information Administration (NTIA)

The National Institutes of Health (NIH) participates in some interagency working group activities, as well.

FDA shares regulatory responsibilities for wireless phones with the Federal Communications Commission (FCC). All phones that are sold in the United States must comply with FCC safety guidelines that limit RF exposure. FCC relies on FDA and other health agencies for safety questions about wireless phones.

FCC also regulates the base stations that the wireless phone networks rely upon. While these base stations operate at higher power than do the wireless phones themselves, the RF exposures that people get from these base stations are typically thousands of times lower than those they can get from wireless phones. Base stations are thus not the primary subject of the safety questions discussed in this document.

What Kinds of Phones Are the Subject of this Update?

The term "wireless phone" refers here to hand-held wireless phones with built-in antennas, often called "cell," "mobile," or "PCS" phones. These types of wireless phones can expose the user to measurable radiofrequency energy (RF) because of the short distance between the phone and the user's head. These RF exposures are limited by Federal Communications Commission safety guidelines that were developed with the advice of FDA and other federal health and safety agencies. When the phone is located at greater distances from the user, the exposure to RF is drastically lower because a person's RF exposure decreases rapidly with increasing distance from the source. The so-called "cordless phones," which have a base unit connected to the telephone wiring in a house, typically operate at far lower power levels, and thus produce RF exposures well within the FCC's compliance limits.

What Are the Results of the Research Done Already?

The research done thus far has produced conflicting results, and many studies have suffered from flaws in their research methods. Animal experiments investigating the effects of radiofrequency energy (RF) exposures characteristic of wireless phones have yielded conflicting results that often cannot be repeated in other laboratories. A few animal studies, however, have suggested that low levels of RF could accelerate the development of cancer in laboratory animals. However, many of the studies that showed increased tumor development used animals that had been genetically engineered or treated with cancer-causing chemicals so as to be pre-disposed to develop cancer in the absence of RF exposure. Other studies exposed the animals to RF for up to 22 hours per day. These conditions are not similar to the conditions under which people use wireless phones, so we don't know with certainty what the results of such studies mean for human health.

Three large epidemiology studies have been published since December 2000. Between them, the studies investigated any possible association between the use of wireless phones and primary brain cancer, glioma, meningioma, or acoustic neuroma, tumors of the brain or salivary gland, leukemia, or other cancers. None of the studies demonstrated the existence of any harmful health effects from wireless phone RF exposures. However, none of the studies can answer questions about long-term exposures, since the average period of phone use in these studies was around three years.

What Steps Can I Take to Reduce My Exposure to Radiofrequency Energy from My Wireless Phone?

If there is a risk from these products—and at this point we do not know that there is—it is probably very small. But if you are concerned about avoiding even potential risks, you can take a few simple steps to minimize your exposure to radiofrequency energy (RF). Since time is a key factor in how much exposure a person receives, reducing the amount of time spent using a wireless phone will reduce RF exposure.

If you must conduct extended conversations by wireless phone every day, you could place more distance between your body and the source of the RF, since the exposure level drops off dramatically with distance. For example, you could use a headset and carry the wireless phone away from your body or use a wireless phone connected to a remote antenna.

Again, the scientific data do not demonstrate that wireless phones are harmful. But if you are concerned about the RF exposure from these products, you can use measures like those described above to reduce your RF exposure from wireless phone use.

Do Hands-Free Kits for Wireless Phones Reduce Risks from Exposure to RF Emissions?

Since there are no known risks from exposure to RF emissions from wireless phones, there is no reason to believe that hands-free kits reduce risks. Hands-free kits can be used with wireless phones for convenience and comfort. These systems reduce the absorption of RF energy in the head because the phone, which is the source of the RF emissions, will not be placed against the head. On the other hand, if the phone is mounted against the waist or other part of the body during use, then that part of the body will absorb more RF energy. Wireless phones marketed in the U.S. are required to meet safety requirements regardless of whether they are used against the head or against the body. Either configuration should result in compliance with the safety limit.

Do Wireless Phone Accessories That Claim to Shield the Head from RF Radiation Work?

Since there are no known risks from exposure to RF emissions from wireless phones, there is no reason to believe that accessories that claim to shield the head from those emissions reduce risks. Some products that claim to shield the user from RF absorption use special phone cases, while others involve nothing more than a metallic accessory attached to the phone. Studies have shown that these products generally do not work as advertised. Unlike "hand-free" kits, these so-called "shields" may interfere with proper operation of the phone. The phone may be forced to boost its power to compensate, leading to an increase in RF absorption. In February 2002, the Federal Trade Commission (FTC) charged two companies that sold devices that claimed to protect wireless phone users from radiation with making false and unsubstantiated claims. According to FTC, these defendants lacked a reasonable basis to substantiate their claim.

What about Children Using Wireless Phones?

The scientific evidence does not show a danger to users of wireless phones, including children and teenagers. If you want to take steps to lower exposure to radiofrequency energy (RF), the measures described above would apply to children and teenagers using wireless phones. Reducing the time of wireless phone use and increasing the distance between the user and the RF source will reduce RF exposure.

Some groups sponsored by other national governments have advised that children be discouraged from using wireless phones at all. For example, the government in the United Kingdom distributed leaflets containing such a recommendation in December 2000. They noted that no evidence exists that using a wireless phone causes brain tumors or other ill effects. Their recommendation to limit wireless phone use by children was strictly precautionary; it was not based on scientific evidence that any health hazard exists.

What about Wireless Phone Interference with Medical Equipment?

Radiofrequency energy (RF) from wireless phones can interact with some electronic devices. For this reason, FDA helped develop a detailed test method to measure electromagnetic interference (EMI) of implanted cardiac pacemakers and defibrillators from wireless telephones. This test method is now part of a standard sponsored by the Association for the Advancement of Medical instrumentation (AAMI). The final draft, a joint effort by FDA, medical device manufacturers, and many other groups, was completed in late 2000. This standard will allow manufacturers to ensure that cardiac pacemakers and defibrillators are safe from wireless phone EMI.

FDA has tested hearing aids for interference from handheld wireless phones and helped develop a voluntary standard sponsored by the Institute of Electrical and Electronic Engineers (IEEE). This standard specifies test methods and performance requirements for hearing aids and wireless phones so that no interference occurs when a person uses a "compatible" phone and a "compatible" hearing aid at the same time. This standard was approved by the IEEE in 2000.

FDA continues to monitor the use of wireless phones for possible interactions with other medical devices. Should harmful interference be found to occur, FDA will conduct testing to assess the interference and work to resolve the problem.

Part Two

Security Issues and Your Telephone Company

Chapter 15

What Your Telephone Company Knows about You

What Does Your Telephone Company Know about You?

Your local, long distance or cellular telephone company knows what numbers you call, how often you call them, how much you pay to call them, what services you subscribe to, how you use those services, and other personal and sensitive information about your telephone usage. This information is called Customer Proprietary Network Information (CPNI). Under the law, telephone companies have a duty to protect this information.

How Can Telephone Companies Use this Information?

A telephone company must obtain customer approval to use, or to share with its affiliates, CPNI to market to the customer services and products that the customer does not already receive from that company. A telephone company may use CPNI to market to the customer services and products that the customer currently receives from that company without additional approval from the customer.

"What Your Telephone Company Knows About You (And Controlling How They Use It)," Federal Communications Commission (FCC), http://www.fcc.gov/cgb/consumerfacts/phoneaboutyou.html, reviewed and updated February 8, 2002.

How Can Telephone Companies Obtain Customer Approval to Use This Information?

There are two ways companies obtain customer approval. One is by sending the customer a notice telling him or her that the company will use (and/or share with its affiliates) his or her CPNI to market products and services that the customer does not currently subscribe to—unless the customer tells the company not to do so. This is known as the “opt-out” method, because the customer’s approval is assumed unless he or she “opts-out” of the company’s use of the CPNI.

The other method is known as the “opt-in” method. Under this method, the company will not use, or share with its affiliates, the customer’s CPNI to market to the customer products and services that the customer does not currently subscribe to, unless the customer expressly gives the company permission to do so. In that way, the customer “opts-in” to the company’s use of his or her CPNI.

How Can I Control the Way Telephone Companies Use This Information?

- Read your telephone bill and any other notices you receive from your telephone company
- Determine if your company is using the opt-in or opt-out method
- Decide if you want your telephone company to use, or to share with its affiliates, your CPNI to market to you services and products that you do not already receive from your telephone company
- Make your choice clear to your telephone company

Remember: These rules apply to all telephone companies: local, long distance and cellular. You will have to make your decision known to each company about how you want it to use your CPNI.

Chapter 16

Slamming: When Your Telephone Service Is Switched without Your Permission

"Slamming" is the illegal practice of changing a consumer's telephone service without permission. New consumer protection rules created by the Federal Communications Commission (FCC) provide a remedy if you've been slammed.

Your Rights If You Have Been Slammed

If You Have Been Slammed and HAVE NOT Paid the Bill of the Carrier Who Slammed You

You DO NOT have to pay anyone for service for up to 30 days after being slammed. This means you do not have to pay either your authorized telephone company (the company you actually chose to provide service) or the slamming company. You must pay any charges for service beyond 30 days to your authorized company, but at that company's rates, not the slammer's rates.

If You HAVE Paid Your Phone Bill and Then Discover That You Have Been Slammed

The slamming company must pay your authorized company 150% of the charges it received from you. Out of this amount, your authorized company will then reimburse you 50% of the charges you paid

---

"When Your Telephone Service Is Switched without Your Permission—Slamming," Federal Communications Commission (FCC), http://www.fcc.gov/cgb/consumerfacts/slamming.html, updated on March 25, 2002.

to the slammer. For example, if you were charged $100 by the slamming company, that company will have to give your authorized company $150, and you will receive $50 as a reimbursement.

With these rules, the FCC has taken the profit out of slamming and protected consumers from illegal charges.

New Guidelines for Telemarketing Switches

Before a telephone company can place an order to switch a customer who agreed to sign up during a telemarketing call, the company must use at least one of the following methods to verify that the customer authorized the switch:

- Obtain a written or electronic Letter of Agency (LOA) from the customer. Any written or electronic LOA used to confirm a telemarketing order must include:
  1. the subscriber's billing name and address
  2. each telephone number to be covered by the order to change the subscriber's telephone company
  3. a statement that the subscriber intends to change from his or her current telephone company to this new company
  4. a statement that the subscriber designates this new carrier to act as the agent for this change, and
  5. a statement that the subscriber understands that there may be a charge for this change. It must also be separate from any promotional material—like prizes, contests, and forms—that come with it.

NOTE: The LOA provided by the carrier must be limited strictly to authorizing a change in telephone carrier and it must be clearly identified as an LOA authorizing the change. The LOA must be written in clear language and the print must be of sufficient size and readable style, generally comparable in type style and size to any promotional materials, and must make clear to the consumer that the document, when signed, would change his or her telephone carrier.

Only the name of the telephone carrier that will set the consumer's rates can appear on the letter of authorization. The LOA must also contain full translations if it uses more than one language.

Advertising promotions that use checks can incorporate an LOA but must meet specific guidelines. A check must contain the necessary information to make it a negotiable instrument and shall not contain any other promotional language or material. The carrier must place the required LOA language near the signature line on the back of the check. In addition, the carrier must print on the front of the check, in easily readable, bold-faced type, a notice that the consumer's signature will authorize a change in his or her telephone carrier.

- Provide a toll-free number that the consumer can call to confirm the order to switch telephone companies.
- Have an independent third party verify the customer's authorization to switch.

NOTE: The Communications Act makes telephone companies responsible for the acts of their agents, including their telemarketers.

How to Avoid Being Slammed

Be a Smart Consumer:

- Always examine your phone bill immediately and thoroughly.
- Be aware of the ways in which companies are legally permitted to change your telephone service. The FCC's rules require companies to obtain your clear permission before such a change. For example, a company may send you an LOA to verify that you want to switch your service to a new company. The LOA is only valid if you sign and date it. It must be used solely to authorize a change in company, and it must be clearly identified as an LOA authorizing the change. Only sign it when you are sure you want to change companies.
- A company might also solicit your telephone business over the phone or electronically. Companies must then verify your authorization by asking you to confirm your order by some means, such as calling a toll-free number used exclusively for this purpose. A company may also employ an independent third party to verify your request to change telephone companies.

What to Do if You've Been Slammed

If Your Telephone Company Has Been Changed without Your Permission:

- Call the slamming company and tell them that you want the problem fixed. If you have not paid, tell them that you will not pay for the first 30 days of service. Call the authorized company (local or long distance) to inform them of the slam. Tell them that you want to be reinstated to the same calling plan you had before the slam.

Tell them that you want all "change of carrier charges" (charges for switching companies) removed from your bill.

- You can also file a complaint. Depending on where you live, you will either file with your state or with the FCC. You can find out whether or not your state will accept complaints by checking the FCC Web site at www.fcc.gov/cgb. You can also check with your state's regulatory commission or Attorney General. The number for your state's regulatory commission, Attorney General, or Consumer Affairs Office is in the blue pages (the "State Government" section) of your phone book. Your state's regulatory commission or Attorney General's Office can advise you on the appropriate procedures for filing complaints with local authorities. In addition, the FCC's Consumer Center at 1-888-CALL-FCC (1-888-225-5322) voice, or 1-888-TELL-FCC (1-888-835-5322) TTY, provides information on slamming and slamming complaints. If your state does not handle slamming complaints, contact the FCC at these numbers for instructions on how to file a complaint with the FCC.

Chapter 17

Cramming: Unauthorized, Misleading or Deceptive Charges Placed on Consumers' Telephone Bills

Background

"Cramming" is the practice of placing unauthorized, misleading, or deceptive charges on your telephone bill. Entities that fraudulently cram people appear to rely largely on confusing telephone bills in order to mislead consumers into paying for services that they did not authorize or receive.

In addition to providing local telephone service, local telephone companies often bill their customers for long distance and other services that other companies provide. When the local company, the long distance telephone company, or another type of service provider either accidentally or intentionally sends inaccurate billing data to be included on the consumer's local telephone bill, cramming can occur.

Cramming also occurs when a local or long distance company or another type of service provider does not clearly or accurately describe all of the relevant charges to the consumer when marketing the service. Although the consumer did authorize the service, the charge is still considered "cramming" because the consumer was misled.

Cramming Charges: What They Look Like

Cramming comes in many forms and is often hard to detect unless you closely review your telephone bill. The following charges

---

"Unauthorized, Misleading, or Deceptive Charges Placed on Your Telephone Bill—'Cramming'," Federal Communications Commission (FCC), http://www.fcc.gov/cgb/consumerfacts/cramming.html, updated February 12, 2002.

would be legitimate if a consumer had authorized them but, if unauthorized, these charges could constitute cramming:

- Charges for services that are explained on a consumer's telephone bill in general terms—such as "service fee," "service charge," "other fees," "voicemail," "mail server," "calling plan," "psychic," and "membership;"
- Charges that are added to a consumer's telephone bill every month without a clear explanation of the services provided—such as a "monthly fee" or "minimum monthly usage fee;" and
- Other charges from a local or long distance company for a service that it provides but, like the other examples, could be cramming if unauthorized.

While cramming charges typically appear on consumers' local telephone bills, they may also be included with bills issued by long distance telephone companies and companies providing other types of services, including cellular telephone, digital telephone, beeper and pager services.

The FCC's Truth-in-Billing Rules

The Federal Communications Commission (FCC) has rules that require telephone companies to make their phone bills more consumer-friendly. These rules enable consumers to more easily determine, when reading their bills, what services have been provided, by whom, and the charges assessed for these services. Telephone companies must also list a toll-free number on their bills for customers with billing inquiries.

Such basic information empowers consumers to protect themselves from cramming and other types of telecommunications fraud. It also helps consumers make informed choices when they shop around to find the best telephone service to meet their needs.

How to Protect Yourself and Save Money

- Carefully review your phone bill every month. Treat your telephone service like any other major consumer purchase or service. Review your monthly bills just as closely as you review your monthly credit card and bank statements.
- Ask yourself the following questions as you review your telephone bill:

1. Do I recognize the names of all the companies listed on my bill?
2. What services were provided by the listed companies?
3. Does my bill include charges for calls I did not place and services I did not authorize?
4. Are the rates and line items consistent with the rates and line items that the company quoted to me?

- You may be billed for a call you placed or a service you used, but the description listed on your telephone bill for the call or service may be unclear. If you don't know what service was provided for a charge listed on your bill, ask the company that billed the charge to explain the service provided before paying the charge.
- Make sure you know what service was provided, even for small charges. Crammers often try to go undetected by submitting $2.00 or $3.00 charges to thousands of consumers.
- Keep a record of the telephone services you have authorized and used—including calls placed to 900 numbers and other types of telephone information services. These records can be helpful when billing descriptions are unclear.
- Carefully read all forms and promotional materials—including the fine print—before signing up for telephone services or other services to be billed on your phone bill.
- Companies compete for your telephone business. Use your buying power wisely and shop around. If you think that a company's charges are too high or that their services do not meet your needs, contact other companies and try to get a better deal.

Actions You Can Take If You Think You've Been Crammed

Take the following actions if unknown charges are listed on your telephone bill:

- Immediately call the company that charged you for calls you did not place, or charged you for services you did not authorize or use. Ask the company to explain the charges. Request an adjustment to your bill for any incorrect charges.

- Call your own local telephone company. FCC rules require telephone companies to place a toll-free number on their bills for customers to contact with billing inquiries. Explain your concerns about the charges and ask your local telephone company the procedure for removing incorrect charges from your bill.
- If neither the local phone company nor the company in question will remove incorrect charges from your telephone bill, you can file a complaint with the regulatory agency that handles your particular area of concern:
  1. For charges on your telephone bill for non-telephone-related services, your complaint should be filed with the Federal Trade Commission (FTC). Call 1-877-FTC-HELP, or use the FTC's online complaint form at https://rn.ftc.gov/dod/wsolcq$.startup. (An example of non-telephone related services is "content" services such as psychic hotlines.)
  2. For charges for telephone-related services provided within your state, you should contact your state regulatory commission. This information may be listed in the government section of your telephone directory.
  3. For charges related to telephone services between two states or internationally, you should contact the FCC. Complaints about these issues may be filed with the FCC in writing, by phone or by e-mail. [See the "Additional Help and Information" section of this *Sourcebook* for contact information.]

Chapter 18

Surcharges for International Calls to Wireless Phones

Consumers should be aware that placing an international long distance call from your wireline telephone here in the United States to a wireless phone in another country may result in a "surcharge" on your bill in addition to your usual charges.

This can happen because many foreign countries use a "calling party pays" framework. Under a "calling party pays" framework, wireless phone subscribers pay only for the outgoing calls they place to others. The "calling party" must pay for calls placed to wireless phones. As a result, when wireline U.S. customers call foreign wireless customers, foreign carriers may pass through to the U.S. carrier the additional cost of connecting the wireless call. The U.S. carrier may then pass this cost through to the U.S. customer as a surcharge on his or her bill. Examples of some of the highest **per minute** surcharges currently reflected on certain long distance carriers' websites include: $0.22 for the United Kingdom; $0.28 for France; $0.32 for Panama; and $0.33 for Uruguay.

Steps You Can Take

- Check with your long distance carrier for more specific information about international wireless surcharges and for international rates.

"Surcharges for International Calls to Wireless Phones," Federal Communications Commission (FCC), http://www.fcc.gov/cgb/consumerfacts/surcharge.html. Updated September 23, 2002.

- Check your carrier's website, which may list surcharges for calls to particular countries.
- Some countries use unique telephone numbers for wireless telephones. Your carrier may be able to provide those numbers (including on its website) so that you will know in advance whether you are about to incur a surcharge in calling a foreign number.

Chapter 19

Understanding Phone Cards: Usage, Problems, and Complaints

What Is a Pre-Paid Phone Card?

A pre-paid phone card is a card you purchase (for a set price) and use to make long distance phone calls. These cards are usually sold in dollar amounts or by number of minutes.

Why Do People Buy Pre-Paid Phone Cards?

Many people use a pre-paid phone card because of the card's convenience—it can be used anywhere and, since you pay in advance, there is no bill. Pre-paid phone cards are popular among travelers, students, people who frequently call overseas, and those who haven't selected a long-distance service. In addition, pre-paid phone cards are sold in convenient places, such as newsstands, post offices, and stores.

What about International Calls?

Rates for international calls can vary dramatically, based on the country that you call or the way that you make the call. Pre-paid phone cards often offer rates that are much lower than a telephone company's basic international rates.

"Pre-Paid Phone Cards: What Consumers Should Know," Federal Communications Commission (FCC), http://www.fcc.gov/cgb/consumerfacts/prepaidcards.html. Updated August 5, 2002.

How Do I Use a Pre-Paid Phone Card?

A toll-free access phone number and a personal identification number (PIN) are usually printed on each phone card. To make a phone call, you dial the access number and then enter the PIN. An automated voice will ask you to enter the phone number you are trying to call, and it will tell you how much time you have left on your card. It might also give you other information/options.

Phone card companies keep track of how much of a card's calling time is used by the card's PIN number. You can add time to some pre-paid phone cards, and the added cost can usually be billed to a credit card. If you cannot add time to your card, you will need to buy a new one once all the time has been used. Also, pre-paid phone cards often have expiration dates. Make sure to keep track of the date your card expires so you don't lose unused minutes.

Who Makes Your Phone Card Work?

- Carriers are responsible for the telephone lines that carry calls.
- Resellers buy telephone minutes from the carriers.
- Issuers set the card rates and provide toll-free customer service and access numbers.
- Distributors sell the cards to the retailers.
- Retailers sell the cards to consumers (though it is important to remember that a store may not have control over the quality of the card or the service it provides).

What Are Common Complaints about Pre-Paid Phone Cards?

As pre-paid phone cards are increasing in popularity, some common complaints are becoming evident. They are:

- Access numbers and/or PINs (personal identification numbers) that don't work;
- Service or access numbers that are always busy;
- Card issuers that go out of business, leaving people with useless cards;
- Rates that are higher than advertised, or hidden charges;

- Cards that charge you even when your call does not go through;
- Poor quality connections; and
- Cards that expire without the purchaser's knowledge.

How Can I Avoid the Problems Associated with Pre-Paid Phone Cards?

Make sure you understand the rates for your particular phone card. Also check the expiration date, look for a toll-free customer service number provided with or on the card, and make sure you understand the instructions on how to use the card. You may also want to ask your friends and family to recommend cards they have used and liked.

What Should I Do If My Pre-Paid Phone Card Doesn't Work?

First, try calling the customer service number provided with the card. If that doesn't work, call or write your local Consumer Affairs Department or state Attorney General. (These phone numbers are often found in the blue pages of your telephone book.) You can also file a complaint or research the company through your local Better Business Bureau, or contact the Federal Trade Commission (FTC). To contact the FTC, call 1-877-FTC-HELP (1-877-382-4357).

Chapter 20

Caller ID Service: Your Rights under the Law

What Is "Caller ID"?

Caller Identification or "Caller ID" acts like an electronic peephole. It allows a person receiving a phone call to see who is calling before answering the phone. A caller's number and/or name is displayed either on your phone (if your phone has this feature) or on a display unit that you must buy separately. The number and/or name will appear on the display unit or on your phone after the first ring. As the caller, this service lets you identify yourself to the person you are calling.

Caller ID is an optional service offered by telephone companies for an additional monthly fee. (Fees vary by phone company.) You can call your local telephone service provider to find out what this fee is or to obtain additional information about Caller ID. Since the time that Caller ID was first made available, it has been expanded to offer Caller ID on Call Waiting (CIDCW) as well. With CIDCW the call waiting tone is heard and the identification of the second call is seen on the display unit.

Blocking and Unblocking?

The Federal Communications Commission's (FCC) national Caller ID rules protect the privacy of the person called and the person calling

---

"Caller ID," Federal Communications Commission (FCC), http://www.fcc.gov/cgb/consumerfacts/callerid.html, updated July 3, 2002.

by requiring telephone companies to make available free, simple, and uniform per-line blocking and unblocking processes. These rules give a caller the choice of delivering or blocking their telephone number for any interstate (between states) call they make. (The FCC does not regulate intrastate calls.)

- **Per-call blocking**—To block your phone number and name from appearing on a recipient's Caller ID unit on a single phone call, dial ***-6-7** before dialing the phone number. Your number will not be sent to the other party. You must redial *-6-7 each time you place a new call.
- **Per-line blocking**—Some states allow customers to select per-line blocking. With this option, your telephone number will be blocked for every call you make on a specific line—unless you use the per-line unblocking option. If you want your number to be transmitted to the called party, dial ***-8-2** before you dial the number you are calling. You must re-dial *-8-2 each time you place a call.
- **800 number/toll-free calls**—Simply requesting privacy when you call 800, 888, 877 and 866 numbers may or may not prevent the display of your telephone number. When you dial a toll-free number, the party you are calling pays for the call. The called party is able to identify your telephone number using a telephone network technology called Automatic Number Identification. FCC rules limit parties that own toll-free numbers from distributing and using this information, and require phone companies to inform customers that their telephone numbers are being transmitted to toll-free numbers in this way.
- **Emergency services**—Calls to emergency lines are exempted from federal Caller ID rules. State rules and policies govern carriers' obligations to honor caller privacy requests to emergency numbers.
- **Blocking the caller's name**—Some Caller ID services also transmit the name of the calling party. The FCC's Caller ID rules require that when a caller requests his/her number be concealed, a carrier may not reveal the caller's name, either.

Caller ID Quick Tips

Look before you dial:

- To block your telephone number for any call, dial ***-6-7** before dialing the telephone number.
- To unblock your number for any call (if you have a blocked line), dial ***-8-2** before dialing the telephone number.

To learn more about Caller ID or the availability of this service in your area, contact your local telephone company or your state public utility commission.

Part Three

Long Distance Telephone Scams

Chapter 21

How Long Distance Scams Work

You've just sat down to a nice family dinner and what happens? A telemarketer calls and asks you if you want to save big money on your long-distance bills, to which your response is probably one of the following:

- To immediately hang up.
- To say, "Of course not—I am always trying to find ways to spend all of this money that keeps piling up at my house! Paying too much for long distance is one of the best methods I have found to get rid of it!"
- To listen and hear about the deal—How much money are they going to offer you to switch? What is the price per minute? What is the monthly fee?

If you do actually listen to the deal, can you even trust what they tell you? Are they giving you the details you need to make a good decision?

What Makes up Your Long-Distance Charges?

One of the most confusing things about your long-distance phone service is the **fees** that are charged in addition to the actual charge for your

"How Long-Distance Scams Work," How Stuff Works, Inc., http://www.howstuffworks.com/phone-scam.htm. © 2002 How Stuff Works, Inc. Reprinted with permission.

calls. Understanding these charges and how they are calculated will help you avoid paying more than you should be for your long distance.

For example, you may sign up for a $0.07-cents-a-minute long-distance rate and be pretty pleased with your negotiation skills, but is that really what you'll be paying? Did you read all of the fine print? Did you ask the right questions? There are many things that apply when your long-distance charges are computed. Here are a few things to make sure you ask about and understand.

In-State Verses out-of-State Calls

You get a great offer of $0.05/minute from long-distance carrier, but do you realize that the rates that are advertised are typically the rates for state-to-state calls only? The long-distance carrier may make that fact obvious, but more often than not this fact is obscured. You have to read the fine print and ask specifically what long-distance calls within your own state will cost. In-state calls are usually anywhere from $0.08 to $0.13 per minute from most carriers. A few may have better deals though, so if most of your calls are made to locations within your state, make sure you ask specifically about in-state rates.

Monthly Service Fees

You got a $0.07-per-minute rate and feel pretty good about it. However, there's a $4.95-per-month service charge that you have to pay to get the rate. Is that a good deal? Maybe, maybe not. The important question: How many calls do you make? Let's look at an example.

Let's say that you average 150 minutes of long-distance yakking per month. If you are getting a $0.07-per-minute rate with a $4.95 service charge, then you're actually paying a little over $0.10 per minute:

$.07 (x) 150 + $4.95 = $15.45 / 150 = **10.45 cents per minute**

Of course, the more you talk, the more you save... or, at least the lower your per-minute cost will be. The key is knowing your calling habits.

Per-Minute Fees

Other aspects that can affect that per-minute fee are things like minimum call length, minimum total long-distance charges, time-of-day rates and billing increments. Let's see how those work.

Minimum Call Lengths

Let's say you get an advertisement for a long-distance service that says you pay $0.10 per minute for calls. You assume (as would most of us) that this means a three-minute call would cost $0.30. But with some carriers, there is a minimum call length in order to get the quoted per-minute rate. You might be told you get $0.10 per minute, but in the fine print it states that there is a $0.50 minimum-call charge. If you get someone's answering machine and hang up after leaving a 20-second message, you're still paying $0.50 for the call.

Minimum Monthly Amounts

You may even have a minimum amount you have to pay in calls each month. For example, some long-distance carriers require a minimum in long-distance calls of anywhere from $20 to $30 per month. You pay this amount even if you don't make ANY long-distance calls.

Time-of-Day Rates

Typically, long-distance carriers break up the day into two rate periods. From 7 in the morning until 7 in the evening is considered the "**peak**" period and gets the highest charged rates. From 7 in the evening until 7 in the morning is considered "**off-peak**" and gets the lowest charged rates. There may also be different rates for **weekends** and **weekdays**. Just because they say $0.07 a minute doesn't mean you'll get that rate around the clock. This might vary with long-distance carriers, so make sure you know when the best times to call are.

Call Billing Increments

How long is a minute, anyway? Sixty seconds? Sometimes it is, sometimes it isn't. Check with your long-distance carrier to see what billing increment they use for calculating the length of your long-distance calls. If it's a 60-second interval, then a second minute is billed if you talk for 61 seconds. If it's a 30-second interval, then one minute is charged if you talk for 31 seconds. If it's a six-second interval, then a full minute is billed only after you've talked for 55 seconds. (If you are connected for 30 seconds, you're only billed for half a minute.) From this, you can see that it is obviously better to look for a smaller increment.

Fees and Taxes

Your long-distance carrier also has the option of passing on some fees to you. We talk about these fees later in this article, but here are some that are linked to your long-distance bill and are not regulated in the amount that can be charged to you. Make sure you check with the long-distance carrier to see what it charges its customers for these fees.

- The **Presubscribed Interexchange Carrier Charge** (PICC) is a charge that the long-distance carrier has to pay the local telephone company for providing the phone systems to its customers. (PIC—Presubscribed Interexchange Carrier—is the term the industry uses to refer to a long-distance carrier.) The PICC is usually $1.04 per month, but there is no law about how much it should be, or if it has to be charged at all. Some long-distance carriers don't charge the fee to their customers at all, and some more than double the fee.
- The **Universal Service Fund Charge** (USF) is another charge that you may or may not see on your long-distance bill. This charge, which is used to help provide phone service to rural areas, low-income customers and others, is usually a percentage of your out-of-state and international long-distance charges. The typical amount that long-distance carriers charge is 6 percent, but once again, they can charge anything they want.

Some Well-Known Scams

Let's visit some of the most well-known long-distance scams. Maybe "scam" is too strong of a word—let's call them "purposefully confusing" deals. The source of discomfort is the hidden restrictions that long-distance companies often apply to boost their rates. Unless you read the fine print, and ask questions when it says "ask about restrictions," you may feel as if you've been scammed, but the company is protected because the information was there (you just didn't see it). If, however, a company blatantly makes a statement that goes against the actual offering, you may have a case for the FCC.

Some of the techniques long-distance carriers use to lure you over to their services actually are good deals. You just have to make sure you find the good ones. Watch out for:

- Companies sending you a **check in the mail** to encourage you to switch—You can cash the check, and in doing so it automatically switches you to their service whether you call them or not.

However, rather than signing up for a plan that gives you a great rate, you get signed up for the basic plan that charges you as much as $0.25 a minute. You have to call to get the advertised plan.

- Companies using **old competitors' rates** in their comparison charts—Make sure you verify the rates they post for competitors to make sure those competitors aren't offering a better deal now.
- **Getting a free prepaid calling card** that automatically switches you over to their long-distance service if you use it—Sometimes you do have to call to establish the service and get a code to use with your card, but not always.
- Companies offering **promotional rates**—Are the rates good forever, or will you be charged a higher rate after three months? Is there no monthly fee forever, or will it kick in after the promotional period is over?
- The **minimum call length** that we talked about in the last section—You may get a great rate for calls over the minimum length, but if you don't reach the minimum you'll be paying very high rates. You never realize how many short calls you make until you see all of those charges on your phone bill.
- **Time restrictions** for the rate you've been promised—Restrictions may not be obvious. It may be that the rate you get is only good at night. Does that fit your calling pattern? Make sure you know when you'll get the good rate and when a higher rate will kick in.
- Using a special **directory assistance company** that advertises a flat rate for the service with no additional charge to connect the call—This may sound like a great deal if you have a good per-minute rate with your long-distance carrier. What you don't notice, however, is that by calling the special number for the directory-assistance deal, you're approving having the call connected and charged using that carrier's long-distance service, which is billed at a much higher rate than your regular carrier. The text "**basic rates apply**" might be the only statement you see that can tip you off that this is happening. (SOURCE: FCC)

Slamming

Slamming is the practice of a long-distance carrier illegally switching your long-distance service to its own plan without your knowledge. Slamming is the single largest source of complaints filed with the

Federal Communications Commission (FCC), and the practice is still increasing. Most consumers don't even notice that they've been slammed, because they don't look very closely at their phone bills to see whom they are paying.

Slamming can happen without you doing anything, or it can happen when you enter a contest or respond to offers for free gifts. If you don't read the fine print, you may be authorizing a long-distance carrier to switch your service when you sign your name.

To protect yourself from being slammed, call your local phone company and ask for a **"PIC" freeze**. A PIC (Presubscribed Interexchange Carrier) is what the telecom industry calls your long-distance carrier. When you freeze your PIC, you are requesting that your local phone service not change your long-distance carrier unless you specifically call them and authorize it. There is also **password protection** to help ensure that someone else doesn't call to authorize the change.

If you've **already been slammed**, there are some steps you can take to correct the problem.

1. Call your local phone company and tell them you have been switched to another long-distance carrier without authorization. They should be able to switch you back and **should not charge you** for the switching service.

2. Call your long-distance carrier—the one you thought you had, anyway—and request that they **return you to the same plan** you had before. Again, there should be no charge for this.

3. Call the long-distance company that slammed you and tell them to **remove all charges for the past 30 days** from your bill. You will have to pay for calls made more than 30 days prior to your noticing you had been slammed, but you will pay them to your original long-distance carrier at their rates.

4. If the company who slammed you won't remove the charges, then file a complaint with the FCC.

If you notice you've been slammed after you've paid the bill, then the slamming long-distance carrier will have to pay your original long-distance carrier **150 percent** of the bill. You'll be reimbursed that extra 50 percent by your original long-distance carrier.

Some states require that long-distance carriers use **Third Party Verification** (TPV) to verify that the customer is actually requesting that his or her service be switched to another long-distance carrier.

The TPV will call you and ask you to confirm that you wish to switch your service.

You can always periodically check to see who your long-distance carrier is by calling **1-700-555-4141** from your home phone. To find out who your local phone service provider is, call **1-your area code-700-4141**.

Cramming

Cramming involves illegally putting unknown and unexplained charges on your bill in the hope that you won't notice them, or will pay them anyway. This type of fraud can be club memberships, service programs, or simply vague, generic-sounding charges like "service charge." You may never use or receive the service or product, but you are charged for it. How do these charges get on your phone bill? There are a number of ways. Let's take a look:

- **Contest entry forms** can bury their true intent in the fine print that no one reads or could comprehend if they did read. By signing and listing your phone number (so they can call you when you win!), you'll be authorizing them to bill you for their "service" or "membership," which you've probably never seen or heard of.
- **Direct mail sweepstakes** are another way that your name and number can be captured and used in this type of scam. You may get a notice in the mail about a sweepstakes and instructions to call a number to see if you've won. This may automatically enroll you in a service that you don't want, but will be billed for on your phone bill.
- **Free calls via 800 numbers**, like those for psychic services, or adult entertainment, request that you state that you "want the service" in order to get through to the psychic. This statement is recorded and used to enroll you in a service, club, or other type of billed program. You never receive anything—probably not even the psychic reading—and are billed on your phone bill for the "service."
- **Instant calling cards** are access codes that reference your phone number and charge paid calls for adult entertainment to your phone bill. This may come about when a visitor (or family member) uses your phone to call an 800 number for an adult service. The person is given the code to use for future calls, and the code charges back to your phone number, not the caller's.

- **Dating services** may also be a way of scamming unsuspecting people. When you call the service to talk with your date, you are told that your date will call you back and you'll need to enter a code to be teleconferenced. These charges are billed to your phone number and are usually mislabeled as collect or toll charges from other cities. (SOURCE: FCC)

To protect yourself from cramming, always review your phone bill for unknown charges. If you find charges you can't identify, contact your local phone company and find out how to dispute the charge.

- **Request that a 900 number block** be placed on your number.
- **Don't respond** when a recording requests that you state "yes" or "I want the service."
- **Don't enter codes** into your phone unless you know what they are for and know it is a trusted source.
- **Always read the fine print** when you register for a contest or sweepstakes.

If you think you've been crammed, call the company who placed the charges on your bill and try to get an explanation. If they are charges for services or products you did not authorize, ask them to remove them. Then, call your local phone company and report the incident. They should be able to help you get the charges removed. File a complaint with the FCC if you can't get any cooperation. Or, you can contact your state Attorney General's office.

10-10-XXX: Dial-around Services Exposed!

Dial-around services are one of those services you see advertised that tell you to dial "10-10" and then three more numbers to bypass your regular long-distance carrier and save money. What they don't tell you is that you may not save any money at all; in fact, you may spend more for your long distance.

While not every dial-around service is out to scam you, some are, and you should be aware of how they operate. The catch with dial-around services is sometimes hard to figure out. Here are some examples of things to watch out for.

- Many times, the discount or advertised price only kicks in after you've talked for 10 or 20 minutes. Until then, you'll be paying extremely high rates for a short call.

- They may say that they offer a 50-percent savings over "basic" long-distance rates. That may still be higher than your existing long-distance carrier charges because you probably aren't subscribing to the "basic" service.
- They may advertise a 20-minute call for $0.99. What they don't tell you up front is that any call under 20 minutes is $0.99, so even a one-minute call costs $0.99.
- They may advertise $0.10 per minute. What they don't tell you up front is that there is also a $0.10 fee for every call, so even a one-minute call costs $0.20.
- They may advertise $0.09 per minute on evenings and weekends, and $0.20 per minute for weekdays. What they don't tell you up front is that they also charge a 4.8-percent surcharge on all of your calls for the Universal Service Fund (USF), which is probably also charged by your regular long-distance carrier.
- They may advertise $0.05 per minute for the first 60 days, and then $0.07 per minute thereafter. What they don't make obvious is that there is also a $4.95 service fee and $1.49 USF charge each month.

As you can see, you really have to ask questions and read all of the literature about these types of services. They can be a good deal in some instances. Just make sure you find out about:

- **Monthly service fees**.
- **Minimum call lengths**.
- **USF and other surcharges**.
- **Restrictions and rate differences based on the time of day**.

There may also be restrictions from your regular phone service. For example, some wireless phone contracts don't allow dial-arounds or phone-card use.

Calling Cards Beware!

Prepaid Calling Cards

Prepaid calling cards are those cards you see being sold everywhere that give you $5 to $50 (or more) worth of long-distance calls.

You buy the card and use it with any phone to make long-distance calls. The card has a **toll-free access number** that you dial when you want to make a call. You then enter a **Personal Identification Number** (PIN) that activates the card's account in the company's computer system. You'll then be prompted to enter the number you wish to call. It may be a lot of numbers to dial, but if you do your homework you can save money.

The retailers that sell these prepaid cards are just one player in the game. There are actually many players.

- There is the **long-distance carrier** who sells blocks of time to resellers.
- These **resellers** sell to or work with **card issuers** who set the card rates, set up access numbers, provide customer service, and maintain the PIN numbers and account information.
- **Distributors** and **retailers** then sell the cards to their customers.

You won't usually know who the long-distance carrier is behind your card.

The price per minute with prepaid phone cards varies greatly, as do the number of additional fees that might be involved. The card may advertise that you can talk for 120 minutes, but that may be under very specific circumstances. While there are certainly cards that are good deals, and the convenience of the cards is perfect for some situations, there are still a lot of scams out there that take advantage of unsuspecting buyers.

The most common scams involve:

- Unadvertised (or non-obvious) **call connection charges**.
- Unadvertised (or non-obvious) **monthly fees**.
- Unadvertised (or non-obvious) **minimum call lengths**.
- Unadvertised (or non-obvious) **quick expiration dates**.
- Unadvertised (or non-obvious) **activation or setup fees**.

These fees eat into the total dollar value of the card, making that 120 minutes you think you're getting turn into a lot fewer. If you made a single call for 120 minutes, perhaps you would get the advertised deal. Making several shorter calls that each racks up connection fees and builds up higher charges because of the minimum call lengths, however, will result in getting far fewer minutes for the card.

Some other things to watch out for with prepaid phone cards include:

- **Pay-phone fees**—In addition to the per-minute charge, connection fees and any other charges you may be paying, if you make your call from a pay phone you should expect to pay an additional surcharge. While the FCC approves an extra charge, card issuers vary greatly in the amount they charge.
- **Large billing increments**—These can be as high as five-minute increments, meaning that **even calls under one minute** use five minutes of the card's value. Any call less than 10 minutes, but more than five minutes, uses 10 minutes of the card's value.
- **Delivery charges**—If you order your card from the Internet or another source, you may also be charged a delivery charge. You also stand the chance of not receiving your card or not receiving a valid PIN.
- **No quality guarantees**—Look for quality guarantees, as well as contact information for any problems you may have.
- **Invalid PINs**—There have been instances of PINs that don't work.
- **Card issuers that go out of business**—This makes your card useless and gives you nowhere to go for a refund.
- **Minute usage beginning as soon as you dial**—Logically, you would think it would begin when a connection is made, but sometimes it doesn't.
- **Busy access numbers**—If you can't connect to the access number, you'll never get through to make the call in the first place.
- **Higher rates** for calls made to wireless or cellular phones.
- **Instant calling cards**—These services allow you to purchase the card online, get an access number and PIN and begin using it right away. While this can work fine and be a good deal, you may also find that buying a prepaid card this way leads to more problems with PINs not working or cards never being received.

According to the Federal Trade Commission (FTC), there are four questions you should always ask before you buy a prepaid phone card:

1. What is the **connection fee** for each call?
2. Is there a **service fee**?
3. Is there a **maintenance fee**?
4. Is there an **expiration date**?

In addition to these questions, you should make sure that you have a **valid customer service number**.

Call the **International Telecard Association** at 1-800-333-3513 to request an informational brochure.

About Pay Phones

Regardless of the type of card you use, if you use a pay phone you'll be using an **Operator Service Provider** (OSP)—the long-distance service provider for pay phones. Unless you dial a special number to connect to the long-distance carrier of your choice, you'll be billed at that OSP's long-distance rates. AND, even if you use a calling card, if that card includes your phone number as part of the card number, the OSP may be able to bill you for the call at its rates.

The only way to be sure you'll be connected with your chosen long-distance carrier and billed at its rates is to follow the directions on the pay phone to call a different long-distance carrier. These directions are required by the FCC, so every pay phone should have them prominently displayed.

Billed Calling Cards

These are credit cards specifically for making long-distance phone calls. Rather than paying for the calls upfront like with the prepaid calls, you pay on a monthly basis after you make the calls.

You can usually get a calling card through your existing long-distance carrier, or through a different carrier. Some carriers do require, however, that your residential or business's long-distance service be through a specific carrier that they are usually affiliated with. Or, they may be restricted to certain geographic areas. Like your regular long-distance service, you need to carefully examine each carrier's rates, charges, restrictions and everything else we've discussed here.

There might be:

- A minimum monthly amount.
- Different rates for different times of the day.

- Surcharges for pay phones.
- A per-call surcharge.

These types of cards are sometimes considered safer than their prepaid cousins. Since you're not paying for something you haven't received yet, you're not taking as much of a risk. If the service is through your existing long-distance carrier, you're also dealing with a known, rather than an unknown.

How to Cut through the Fluff

So how do you know if you're getting the best deal with all of the rates, fees, tricks and scams that you have to watch out for? Are you up to the challenge of deciphering the ads and information on long-distance carriers' Web sites and comparing their rates, fees and special packages side by side? Don't worry, you don't have to be—SmartPrice.com has already done that. They maintain an unbiased, current database of information about nearly every long-distance carrier.

Their **searchable database** lets you enter some simple information about your calling habits and, in return, get a list of long-distance carriers that can offer you the best deal based on that information. You can compare rates, see all of the additional charges each carrier has, look at their billing increments, see their quality ratings, and switch your service on the spot when you choose to. These services are totally free, and no personal information is necessary to do the search.

What you'll get with the search may not be the lowest per-minute rate, but rather the **lowest overall monthly bill**. The database takes all of those variable fees and charges into consideration. You can then look at each carrier's plan to see which would work best for you.

Using a service like SmartPrice.com simply saves you the time and frustration of trying to, first, find current information about various long-distance carriers, and second, set that information up so that you can compare each rate, fee and other aspect side by side in order to make a good decision. SmartPrice.com was started when one of the co-founders had to do just that. He quickly found that it was next to impossible to get the information he needed to make a good comparison of long-distance carriers.

A Final Note about Your Phone Bill

If you have looked at your phone bill very carefully, you've probably seen a lot of charges that you couldn't identify. Are these charges that you have to pay, or can you somehow get rid of them? Most of them are charges that the government allows local phone companies to charge to recover some of the costs of a local phone network. Let's take a look at what should be there and what you should watch out for.

Here are some of the fees you'll probably see. Keep in mind that some of these have nothing to do with your long-distance service.

- **Municipal Charge**—This is charged to pay for local community services such as 911 and other emergency services.
- **Number Portability Service Charge**—This is an FCC-approved charge that your phone company puts on your bill when you switch long-distance carriers. It pays for the administrative costs of switching from one long-distance carrier to another.
- **Universal Service Charge**—This is a fee that goes into a **Universal Service Fund** that was created to help make phone service available to low-income customers, customers who are in rural areas with higher costs, and customers with disabilities. It is helps pay for Internet access for schools and libraries, and it also helps pay for links for rural health care providers to urban medical centers for advanced diagnostic and other medical services that are available. All telecommunications companies that provide services between states have to contribute to the fund. The fee they pay changes each quarter based on the needs of the fund. They have the option of charging their customers all or a percentage of this fee. Most charge customers based on a percentage of the total bill, a flat fee, or a percentage of just your out-of-state call charges.
- **Federal Subscriber Line Charge**—This is a fee the government allows your local phone company to charge you in order to pay for the telephone lines connected to your home. It isn't a tax that goes to the government, but rather a fee the phone company gets for putting in and maintaining those lines. The government does put a limit on the charge so that phone service stays at an affordable rate for everyone. The charge can be as high as $5.00 for your primary phone line and $7.00 for additional phone lines into your home.
- **Presubscribed Interexchange Carrier Charge**—This is the fee the local phone company charges long-distance carriers for accessing what is called the "local loop." The local loop is all of

the outside wiring, underground conduits, telephone poles and other facilities that are necessary to get phone service into your home and connect you to the network. This charge picks up where the Federal Subscriber Line Charge left off in helping local phone companies pay for the lines, equipment, and their maintenance.

About the Industry

When the telephone service industry was deregulated in 1984, AT&T, who had a monopoly on U.S. telephone service, was broken up into many regional services, known as the "Baby Bells." This all began as a result of MCI (at the time, only a small carrier) suing AT&T over the right to access local telephone exchanges. In 1996, the Telecommunications Act opened up the market for local and long-distance service to even more carriers.

Deregulation has improved the long-distance services you have access to tremendously by creating competition, which leads to better deals. You now not only have more choices, but the choices come with better deals like lower rates, more calling plans and smaller billing increments.

Now, there are over 1,200 long-distance phone services, but you've probably only heard of a handful of them. The largest and most well-known are AT&T, Sprint, MCI WorldCom, and recently Verizon. And there's virtually no difference in the quality of the call, because they all use the same fiber-optic networks. The differences you'll see will be in the operations of the companies, such as the billing practices, the calling packages they offer, their rates and their customer service.

Services are also offered through **Long Distance Resellers**. These are companies that have no facilities for phone service themselves, but purchase blocks of time from large long-distance carriers who do. They buy large blocks of time at a big discount, which allows them to then resell it at a lower price and at the same time make a small profit. Their strategy is to make money through the volume of time they purchase.

This can be a great deal for the consumer because they get the same quality at a lower price.

Chapter 22

The Latest Telephone Scams

Chapter Contents

Section 22.1

Mexico Collect Call Scam: Targeting Hispanics

"Mexico Collect Call Scam," Federal Communications Commission (FCC), http://www.fcc.gov/cgb/consumerfacts/mexicoscam.html. Updated March 25, 2002.

The FCC has learned of a telephone scam that originates in Mexico and apparently targets people in Hispanic communities.

Consumers report that they have been deceived into accepting a collect call from a particular family member when, in fact, the call is from a stranger. The consumer is then fraudulently billed a large amount for a call that lasts a few minutes or less—or for a nonexistent call.

This scam seems to be prevalent in Spanish-speaking communities in California, Texas, Florida, New York and Illinois. The scam is targeted at Spanish-surnamed consumers.

Here's How it Works

An operator calls the consumer's residential telephone number and tells the consumer he/she has a collect call from a family member who has an emergency or an important message. The operator has all the relevant information—the family's last name, husband's name, wife's name, etc. The operator provides the consumer with the "calling family member's" name. The consumer accepts the operator-assisted call, assuming there is a real emergency or message. Upon accepting the call, the consumer is then connected to a complete stranger who gives information that is not related to the consumer's family. Realizing the call is a fraud, the consumer immediately hangs up, but is still billed for the call.

In some cases, the consumer is not even allowed to respond "yes" or "no" in accepting the call; the operator automatically puts the call through without waiting for an affirmative response. Other times, the consumer actually declines the call and is still charged a very high rate for a collect call that was never accepted.

Here's How to Avoid This Scam

Consumers should use voice recognition as a tool for identifying the person placing the collect call. Specifically, consumers should ask the operator to have the person placing the collect call speak his name, instead of allowing the operator to say the name of the person placing the collect call. Also, consumers should carefully examine their monthly telephone bills for accuracy, and report errors to the company billing for the erroneous charges.

Filing a Complaint with the FCC

Consumers who become victims of this scam are encouraged to file a written informal complaint with the FCC. There is no charge for this. [See the "Additional Help and Information" section of this Sourcebook for information on filing a complaint with the FCC.]

Section 22.2

Don't Fall for the 9-0-# Telephone Scam

"Don't Fall for the 9-0-# Telephone Scam," Federal Communications Commission (FCC), http://www.fcc.gov/cgb/consumerfacts/90Scam.html. Updated March 25, 2002.

The old phone scam involving the 9-0-# buttons on your business telephone is still around.

How This Scam Occurs

You receive a call at your office from someone claiming to be a telephone company employee investigating technical problems with your line, or checking up on calls supposedly placed to other states or countries from your line. The caller asks you to aid the investigation by either dialing 9-0-# or transferring him/her to an outside line before hanging up the telephone receiver. By doing this, you may be enabling the caller to place calls that are billed to your office telephone number.

What You Should Know

- Telephone company employees checking for technical and other types of telephone service or billing problems would not call and ask a subscriber to dial a specific series of numbers before hanging up the telephone receiver.
- Telephone company employees would not request subscribers to connect the caller to an outside line before hanging up the receiver.
- These types of calls are made to trick subscribers into taking actions that will enable the caller to place fraudulent calls.
- This scam only works if your telephone is served by a private branch exchange (PBX) or private automatic branch exchange (PABX).

What to Do

If your place of business utilizes either a PBX or a PABX, you or your company telecommunications manager should contact the manufacturer of the PBX or PABX and the telephone companies that provide you with local and long distance service to obtain information about the type of security systems available to protect your telephone system from toll fraud. You may also ask about any monitoring services that help detect unusual telephone system usage.

Avoid Becoming a Target

To avoid becoming a target of this scam, educate yourself and other employees about the 9-0-# scam. Encourage employees to take the following steps if they think that a telephone call is fraudulent or is part of this scam:

- Ask the caller for his/her name and telephone number;
- Tell the caller you are going to call the telephone company immediately to determine whether or not there is a problem with the line;
- Immediately hang up the receiver; do not dial any numbers or transfer the caller to an outside line before hanging up;
- Find the telephone number for your telephone service provider and/or its security office and report the suspicious phone call. Be prepared to provide details of the call to the telephone company representative; and
- Contact your local law enforcement officials.

Chapter 23

Recurring Telephone Scams

Chapter Contents

Section 23.1

900 Number Pay-Per-Call and Other Information Scams

"900 Number Pay-Per-Call and Other Information Scams," Federal Communications Commission (FCC), http://www.fcc.gov/cgb/consumerfacts/900Fact.html. Updated April 25, 2002.

What Are Pay-Per-Call and Information Services?

Although the terms are often used interchangeably, there are differences between "information" and "pay-per-call" services. Information services offer telephone callers the opportunity to obtain a wide variety of recorded or live information and entertainment. For example, information services may provide medical, stock market, sports or product information. Information services can be reached by dialing numbers other than "900" numbers.

Pay-per-call services are a particular type of information service. They are offered through 900 numbers and always carry a fee greater than the cost of simply transmitting the call. The fee may be a per-minute charge or a flat fee. Pay-per-call services can include "adult" services, "chat" lines, and psychic advice.

The Federal Communications Commission (FCC) receives a considerable number of complaints involving information and pay-per-call services.

Consumers call an advertised telephone number to receive a pay-per-call or other information service. Fees are typically charged to the telephone number from which the call was placed and appear on the monthly telephone bill.

Several different entities may be involved in providing lawful information services:

- **Information providers** ("IPs") design, produce, price, and advertise pay-per-call and other information services. They typically use **long distance telephone companies** to transmit their information services to callers.

- **Long distance companies** transmitting 900 number services also usually offer IPs billing and collection services along with pay-per-call transmission. Long distance companies may directly bill consumers for information service calls or may subcontract with local telephone companies to place information service charges on their monthly telephone bills.
- An IP may obtain transmission and billing services from an entity called a **service bureau**. Service bureaus act as middlemen between IPs and long distance companies. Service bureaus may also provide other services to individual IPs, including actual information programs, management services, and billing and collection of charges.
- Finally, an IP may employ an **independent billing company** or **collection agency** to attempt to collect information service charges that either are not billed by a telephone company or have been removed from a phone bill.

FCC Rules

The FCC regulates U.S. telephone companies that are involved in transmitting and billing interstate pay-per-call and other information services. The FCC's rules governing information services provide that:

- Any interstate (between states) service—other than telephone company directory assistance—that charges consumers for information or entertainment must be provided through a 900 number unless it is offered under what is termed a "presubscription or comparable arrangement." That arrangement may be a preexisting contract by which the caller has "subscribed" to the information service. The arrangement also may be the caller's authorization to bill an information service call to a prepaid account or to a credit, debit, charge, or calling card;
- Telephone companies may not disconnect local or long distance telephone service for failure to pay 900 number charges or charges for presubscribed information services listed on the phone bill;
- Local telephone companies must offer consumers the option of blocking access to 900 number services, if technically feasible;
- Telephone companies that bill consumers for pay-per-call and presubscribed information services must show those charges in

a portion of the bill that is separate from local and long distance charges. In addition, telephone companies must include, with the bill, certain information outlining consumers' rights and responsibilities with respect to payment of information service charges;

- Toll-free numbers may not be used to initiate collect calls from IPs; and
- Callers to toll-free numbers may not be transferred to 900 numbers.

Protecting Yourself from Unwanted Charges for Information Services

- If you dial a 900 number, even if you are calling to claim a "free prize," there will be charges. Listen to the introductory message and hang up if you decide you are not interested in the program or do not wish to pay charges.
- Information services are rarely free, even if they are provided over toll-free numbers. If it sounds too good to be true, it probably is.
- Be careful when making long distance calls, accepting collect calls, or accepting unsolicited offers from IPs.
- Not all advertisements for information services disclose the charges.
- Be cautious when calling an information service that advertises "ordinary toll rates apply" or "international toll rates apply." Be aware that the toll rates applied by some telephone companies involved in transmitting information services may be higher than the rates of your own primary long distance company and that international rates—even those of your own long distance carrier—may be more than you might expect.
- Caution children or other individuals who make phone calls from your telephone line about charges for calls to information services.
- Consider obtaining a 900 number block if you do not wish to call any 900 number. Contact your local and local long distance telephone companies about the availability of blocks for international or toll calls if you are concerned about securing your phone line against these calls.

- Examine your telephone bill carefully each month. Calls to 900 numbers must be listed separately on your bill, but a call to an IP in a foreign country may not necessarily be designated as an information services call—it may be billed as an ordinary toll call or a "calling card" call.
- Don't be intimidated when dealing with telephone companies, IPs, or collection agencies. Learn your rights and challenge incorrect or unwarranted charges for information service calls.

Filing a Complaint with the FCC

[See the "Additional Help and Information" section of this Sourcebook for further information on filing a complaint with the FCC.]

Section 23.2

The 809 Area Code Telephone Scam

"809 Phone Scam—Beware," Federal Communications Commission (FCC), http://www.fcc.gov/cgb/consumerfacts/809.html. Updated March 25, 2002.

The Federal Communications Commission (FCC) has become aware of a long distance phone scam that may lead consumers to inadvertently ring up high charges on their phone bills.

The Scam Works Something Like This

- You get an e-mail, voicemail or page telling you to call a phone number with an 809 (or some other three-digit) area code to collect a prize, find out about a sick relative, engage in sex talk, etc.
- You assume you are making a domestic long distance call—as "809" (and other three-digit area codes involved in this scam) appear to be typical three-digit U.S. area codes.
- When you dial the "809" area code, however, you're actually connected to a phone number outside the United States.

- You don't find out about the higher international call rates until you receive your phone bill.

To Minimize the Risk of This Happening to You

- Check any area codes before returning calls.
- If you do not otherwise make international calls, ask your local phone company to block outgoing international calls on your line.

Filing a Complaint with the FCC

There is no charge to file an informal complaint with the FCC. [See the "Additional Help and Information" section of this *Sourcebook* for information on filing a complaint with the FCC.]

Section 23.3

Toll-Free Telephone Number Scams

"Toll-Free Telephone Number Scams," Federal Trade Commission (FTC), http://www.ftc.gov/bcp/conline/pubs/tmarkg/tollfree.htm. May 1997. Downloaded April 16, 2002. Although the date of this document is 1997, readers will find the information contained within it useful.

Calls to 800 and 888 numbers are almost always free, but there are some exceptions. Companies that provide audio entertainment or information services may charge for calls to 800, 888 and other toll-free numbers, but only if they follow the Federal Trade Commission's 900-Number Rule.

This Rule requires a company to ask you to pay for entertainment or information services with a credit card ***or*** to make billing arrangements with you ***before*** they provide the service. If you don't use a credit card, the law says companies also must provide you with a security device, such as a personal identification number (PIN), to prevent other people from using your phone to charge calls to these services.

Presubscription Agreements

For a company to charge you for a call to an 800 or 888 entertainment or information service, it must obtain your agreement to the billing arrangement in advance. The company must tell you all relevant information about the arrangement, including the company's name and address, rates and rate changes, and business telephone number.

The company also must use a security device, like a PIN, to prevent unauthorized charges to your telephone. The "presubscription agreement" must be in place before you reach the entertainment or information provided by the service. If you authorize a company to charge your credit card for an 800 or 888-number call, the company has met the Rule's requirements.

Prohibitions and Unlawful Practices

Certain practices relating to 800 and 888 numbers are prohibited by the 900-Number Rule. For example, a company can't charge you for dialing an 800 or 888 number unless you have entered into a valid presubscription agreement. Also, if you dial an 800 or other toll-free number, the company is prohibited from automatically connecting you to a 900-number service, and from calling you back collect. However, the law allows a company to ***promote*** a 900-number service during the 800-number call, as long as you would have to hang up and dial the 900 number to reach the service.

Some companies break the law by charging improperly for entertainment and information services that you reach by dialing an 800 or 888 number. For example, some services ask you during the course of a call to simply "Press 1" to be charged automatically. Others advertise a service as "free" but then unlawfully charge for calls placed to that service. Still others may charge for calls you place to 800 or 888 numbers by billing you for calls to a different type of service—such as calls to an international number. Some will charge a "monthly club fee" on your phone bill after you call an 800 or 888 number. Other services fail to take adequate precautions to prevent the unauthorized use of your telephone to make these calls; they may charge you for 800-number calls you didn't make or approve.

Minimize Your Risk

Here's how to minimize your risk of unauthorized charges:

- Remember that dialing a number that begins with 888 is just like dialing an 800 number; both are often toll-free, but not always. Companies are prohibited from charging you for calls to these numbers unless they set up a valid presubscription agreement with you first.
- Recognize that not all numbers beginning with "8" are toll-free. For example, the area code 809 serves the Dominican Republic. If you dial this area code, you'll be charged international long distance rates.
- Make sure any 800 or 888 number you call to get entertainment or information that costs money provides security devices—including PINs—before you enter into a presubscription agreement with them.
- Check your phone bill for 800, 888 or unfamiliar charges. Calls to 800 and 888 numbers should be identified. Some may be mislabeled as "long-distance" or "calling card" calls and are easy to overlook.
- Dispute charges on your phone bill for an 800 or 888 number if you don't have a pre-subscription arrangement. Follow the instructions on your billing statement.
- Realize that if the telephone company removes a charge for an 800 or 888-number call, the entertainment or information service provider may try to pursue the charge through a collection agency. If this happens, you may have additional rights under the Fair Debt Collection Practices Act.

For More Information

The following organizations can provide additional information and help you file a complaint.

- Your state Attorney General usually has a division that deals with consumer protection issues.
- The Federal Communications Commission's National Call Center at 1-888-CALL-FCC (1-888-225-5322). The Center answers consumer inquiries relating to communications law and policy, matters pending before the FCC, and any possible violations of FCC law or policy.

Section 23.4

Phone, E-mail, and Pager Messages May Signal Costly Scams

"Phone, E-Mail & Pager Messages May Signal Costly Scams," Federal Trade Commission (FTC), http://www.ftc.gov/bcp/conline/pubs/alerts/phonscam.htm. December, 1996. Downloaded March 10, 2002. Although the date of this document is 1996, readers will find the information contained within it useful.

Beware the "urgent" message you hear on your answering machine, the e-mail message about a "prize," or the long-distance message on your pager. Any of them could turn out to be an expensive telephone trap.

From coast to coast, American consumers are getting stung by "emergency" and cryptic telephone, e-mail and pager messages urging them to call an "809" number for information about injured or sick relatives, "prize opportunities" or "debt collectors." The messages tell recipients to call a long-distance number for more information. In many cases, the return-call number is an international pay-per-call line, with a three digit exchange that looks like an American or Canadian area code.

Concerned or curious consumers who take the bait and place the return call usually are kept on the line, listening to long-winded messages. As the clock ticks, the charges build, and scam artists are counting the rebates they'll receive from foreign telephone companies. For every minute you stay on the line, the scam artist who offers the "information" collects a bigger share in the profits.

It's not always easy to distinguish an international dialing code from a North American area code. Most international numbers can be reached only by dialing 011, the international access code. However, some places outside the United States or Canada, such as the Caribbean, can be reached simply by dialing a number beginning with three digits that resemble a North American area code. Many scam artists take advantage of this situation—and of unsuspecting consumers—by urging them to call numbers that begin with area codes 809, 758, or 664 without revealing that these calls result in international long distance charges that could be costly. Because each country establishes its own telephone rates, there is no limit to the per-minute charge for these calls. The Federal Trade Commission reminds consumers to be suspicious of unidentified telephone, e-mail, or pager messages that claim to offer information about a sick or injured relative, a debt, bad credit, or prize offer. In addition, the FTC cautions consumers to be wary of messages from unfamiliar sources with a return telephone number using 809, 758, or 664 area codes, or the 011 international access code. Finally, the agency suggests that consumers question television or print ads that offer products or investment opportunities through telephone numbers that start with these area codes or the 011 international code.

Table 23.1 Area Code Table

Area Code	Location
264	Anguilla
268	Antigua and Barbuda
242	Bahamas
246	Barbados
441	Bermuda
284	British Virgin Islands
245	Cayman Islands
767	Dominica
473	Grenada
876	Jamaica
664	Montserrat
869	St. Kitts and Nevis
758	St. Lucia
784	St. Vincent/ Grenadines
868	Trinidad and Tobago
809	Dominican Republic

Section 23.5

Reloading Scams: Double Trouble for Consumers

"Reloading Scams: Double Trouble for Consumers," Federal Trade Commission (FTC), http://www.ftc.gov/bcp/conline/pubs/tmarkg/reload.htm. May 1998. Downloaded March 10, 2002. Although the date of this document is 1998, readers will find the information contained within it useful.

If you've been a victim of telemarketing fraud, chances are you're on a list to be called—and scammed—again. That's because consumers who have lost money often are placed on "sucker lists," an index of people who have lost money to bogus telephone solicitations. Sucker lists, which include names, addresses, phone numbers, and other information, are created, bought, and sold by some fraudulent telemarketers. They're considered invaluable because dishonest promoters know that consumers who have been tricked once are likely to be tricked again. This double scam is called "reloading."

How the Scam Works

Double scammers, known as reloaders, use several methods to repeatedly victimize consumers. For example, if you've lost money to a fraudulent telemarketing scheme, you may get a call from someone claiming to work for a government agency, private company, or consumer organization that could recover your lost money, product, or prize—for a fee. The catch is that the second caller often is as phony as the first, and may even work for the company that took your money in the first place. If you pay the recovery fee, you've been double-scammed.

Be aware that some local government agencies and consumer organizations provide help to consumers who have lost money to fraudulent promoters. Fortunately, there's a way to tell whether the caller offering help is legitimate: If they ask you to pay a fee or if they guarantee to get your money back, it's a fraud.

Another reloading scam uses prize incentives to get you to continue to buy merchandise. If you buy, you may get a second call claiming you're eligible to win a more valuable prize. The second caller makes you think that buying more merchandise increases your chances of winning. If you take the bait, you may be called yet again with the same sales pitch. The only difference is that the caller now claims that you're a "grand prize" finalist and, if you buy even more, you could win the "grand prize."

Fraudulent promoters involved in reloading scams want payment as quickly as possible—usually by credit card or a check delivered to them by courier. Often, it takes at least several weeks to get your products and prizes. When you do receive them, you'll probably find that you've overpaid for shoddy goods, and that you didn't win the "grand prize" at all. Unfortunately, your credit card has long since been charged or your check cashed.

Protect Yourself

You can avoid becoming a victim of a reloading scam. Here's how:

- Beware of people who claim to work for companies, consumer organizations, or government agencies that recover money for a fee. Legitimate organizations, such as national, state, and local consumer enforcement agencies and non-profit organizations, like the **National Fraud Information Center (NFIC)** or **Call For Action (CFA)**, do not charge for their services or guarantee results.
- Be suspicious of people who ask you to send money by a courier or who say they will send a courier to your home to pick up your check.
- Before you buy over the phone from someone you don't know, ask for written information about the deal. Also check out the company with your state or local consumer protection agency or the Better Business Bureau. This is not a foolproof way to check on a company, but it is prudent.
- Be wary of promoters who contact you several times and urge you to buy more merchandise to increase your chances of winning valuable prizes.
- Wait until you get and inspect your first order before you buy more.

Section 23.6

Telephone Scams That Target Seniors

"Are You a Target of...Telephone Scams?" Federal Trade Commission (FTC), http://www.ftc.gov/bcp/conline/pubs/tmarkg/target.htm. Undated document; downloaded March 10, 2002.

If you're age 60 or older, you may be a special target for people who sell bogus products and services by phone.

It's easy enough to fall prey to their sales pitch. Telemarketing fraud is a multi-billion dollar business in the United States. Every year, thousands of consumers lose from a few dollars to their life savings to telephone con artists.

That's why the Federal Trade Commission (FTC) encourages you to be skeptical when you hear a phone solicitation and to be aware of the Telemarketing Sales Rule, a new law that can help you protect yourself from abusive and deceptive telemarketers.

How Older People Become Victims of Telemarketing Fraud

Fraudulent telemarketers try to take advantage of older people on the theory that they may be more trusting and polite toward strangers. Older women living alone are special targets of these scam artists.

Here are some reasons older people become victims of telemarketing fraud:

- Often it's hard to know whether a sales call is legitimate. Telephone con artists are skilled at sounding believable—even when they're really telling lies.
- Sometimes telephone con artists reach you when you're feeling lonely. They may call day after day—until you think a friend, not a stranger, is trying to sell you something.
- Some telephone salespeople have an answer for everything. You may find it hard to get them off the phone—even if they're selling something you're not interested in. You don't want to be rude.

- You may be promised free gifts, prizes, or vacations—or the "investment of a lifetime"—but only if you act "right away." It may sound like a really good deal. In fact, telephone con artists are only after your money. Don't give it to them.

Common Telephone Scams

Con artists never run out of scams. Have you heard any of these?

- *Prize offers:* You usually have to do something to get your "free" prize—attend a sales presentation, buy something, or give out a credit card number. The prizes generally are worthless or overpriced.
- *Travel packages:* "Free" or "low-cost" vacations can end up costing a bundle in hidden costs. Or, they may never happen. You may pay a high price for some part of the package—like hotel or airfare. The total cost may run two to three times more than what you'd expect to pay or what you were led to believe.
- *Vitamins and other health products:* The sales pitch also may include a prize offer. This is to entice you to pay hundreds of dollars for products that are worth very little.
- *Investments:* People lose millions of dollars to "get rich quick" schemes that promise high returns with little or no risk. These can include gemstones, rare coins, oil and gas leases, precious metals, art, and other "investment opportunities." As a rule, these are worthless.
- *Charities:* Con artists often label phony charities with names that sound like better-known, reputable organizations. They won't send you written information or wait for you to check them out with watchdog groups.
- *Recovery scams:* If you buy into any of the above scams, you're likely to be called again by someone promising to get your money back. Be careful not to lose more money in this common practice. Even law enforcement officials can't guarantee they'll recover your money.

Tip-offs to Phone Fraud

Telephone con artists spend a lot of time polishing their "lines" to get you to buy. You may hear this:

- You must act "now"—or the offer won't be good.
- You've won a "free" gift, vacation, or prize—but you pay for "postage and handling" or other charges.
- You must send money, give a credit card or bank account number, or have a check picked up by courier—before you've had a chance to consider the offer carefully.
- You don't need to check out the company with anyone—including your family, lawyer, accountant, local Better Business Bureau, or consumer protection agency.
- You don't need any written information about their company or their references.
- You can't afford to miss this "high-profit, no-risk" offer.

If you hear these—or similar—"lines" from a telephone salesperson, just say "no thank you," and hang up the phone.

The Telemarketing Sales Rule

The FTC's Telemarketing Sales Rule requires telemarketers to make certain disclosures and prohibits certain misrepresentations. It gives you the power to stop unwanted telemarketing calls and gives state law enforcement officers the authority to prosecute fraudulent telemarketers who operate across state lines.

The Rule covers most types of telemarketing calls to consumers, including calls to pitch goods, services, "sweepstakes," and prize promotion and investment opportunities. It also applies to calls consumers make in response to postcards or other materials received in the mail.

Keep this information near your telephone. It can help you determine if you're talking with a scam artist or a legitimate telemarketer.

- It's illegal for a telemarketer to call you if you've asked not to be called. If they call back, hang up and report them to your state Attorney General.
- Calling times are restricted to the hours between 8 a.m. and 9 p.m.
- Telemarketers must tell you it's a sales call and who's doing the selling before they make their pitch. If it's a prize promotion, they must tell you that no purchase or payment is necessary to enter or win. If you're asked to pay for a prize, hang up. Free is free.

- It's illegal for telemarketers to misrepresent any information, including facts about their goods or services, earnings potential, profitability, risk or liquidity of an investment, or the nature of a prize in a prize-promotion scheme.
- Telemarketers must tell you the total cost of the products or services they're offering and any restrictions on getting or using them, or that a sale is final or non-refundable, before you pay. In a prize promotion, they must tell you the odds of winning, that no purchase or payment is necessary to win, and any restrictions or conditions of receiving the prize.
- It's illegal for a telemarketer to withdraw money from your checking account without your expressed, verifiable authorization.
- Telemarketers cannot lie to get you to pay, no matter what method of payment you use.
- You do not have to pay for credit repair, recovery room, or advance-fee loan/credit services until these services have been delivered. (Credit repair companies claim that, for a fee, they can change or erase accurate negative information from your credit report. Only time can erase such information. Recovery room operators contact people who have lost money to a previous telemarketing scam and promise that, for a fee or donation to a specified charity, they will recover your lost money, or the product or prize never received from a telemarketer. Advance-fee loans are offered by companies who claim they can guarantee you a loan for a fee, paid in advance. The fee may range from $100 to several hundred dollars.)

Exceptions to the Rule

While most types of telemarketing calls are covered by the Rule, there are exceptions. The Rule does not cover:

- Calls placed by consumers in response to general media advertising, except calls responding to ads for investment opportunities, credit repair services, recovery room services, or advance-fee loans.
- Calls placed by consumers in response to direct mail advertising that discloses all the material information required by the Rule, except calls responding to ads for investment opportunities, prize promotions, credit repair services, recovery room services, or advance-fee loans.

- Catalog sales.
- Calls initiated by the consumer that are not made in response to any solicitation.
- Sales that are not completed, and payment or authorization for payment is not required, until there is a face-to-face sales presentation.
- Calls from one business to another unless nondurable office or cleaning supplies are being offered.
- Sales of pay-per-call services and sales of franchises. These are covered by other FTC rules.

What You Can Do to Protect Yourself

It's very difficult to get your money back if you've been cheated over the phone. Before you buy anything by telephone, remember:

- Don't buy from an unfamiliar company. Legitimate businesses understand that you want more information about their company and are happy to comply.
- Always ask for and wait until you receive written material about any offer or charity. If you get brochures about costly investments, ask someone whose financial advice you trust to review them.
- Always check out unfamiliar companies with your local consumer protection agency, Better Business Bureau, state Attorney General, the National Fraud Information Center, or other watchdog groups. Unfortunately, not all bad businesses can be identified through these organizations.
- Always take your time making a decision.
- Legitimate companies won't pressure you to make a snap decision.
- It's never rude to wait and think about an offer. Be sure to talk over big investments offered by telephone salespeople with a trusted friend, family member, or financial advisor.
- Never respond to an offer you don't understand thoroughly.
- Never send money or give out your credit card or bank account number to unfamiliar companies.

- Be aware that any personal or financial information you provide may be sold to other companies.

For More Help

Before you buy from an unfamiliar organization, check it out with some of these groups. Your local phone directory has phone numbers and addresses. [See the "Additional Help and Information" section of this *Sourcebook* for contact information.]

- Call for Action.
- State Attorney General.
- Better Business Bureau.
- Local consumer protection organization.

To stop unwanted telephone sales calls from many national marketers, send your name, address, and telephone number to the Direct Marketing Association.

Under the Telephone Consumer Protection Act of 1991, you can ask that companies put you on their "do not call" lists. If the company calls you again, you can bring action in Small Claims Court.

Part Four

Telemarketing Fraud and Other Nuisances

Chapter 24

What Is Telemarketing Fraud?

Telemarketing fraud is a term that refers generally to any scheme to defraud in which the persons carrying out the scheme use the telephone as their primary means of communicating with prospective victims and trying to persuade them to send money to the scheme. When it solicits people to buy goods and services, to invest money, or to donate funds to charitable causes, a fraudulent telemarketing fraud operation typically uses numerous false and misleading statements, representations, and promises, for three purposes:

(1) *To make it appear that the good, service, or charitable cause their telemarketers offer to the public is worth the money that they are asking the consumer to send.* Fraudulent telemarketers, by definition, do not want to give consumers fair value for the money they have paid to the telemarketers. Because their object is to maximize their personal profits, even if the consumer suffers substantial financial harm, they will typically adopt one or both of two approaches: to fail to give the consumer anything of value in return for their money; or to provide items of modest value, far below what the consumer had expected the value to be on the basis of the telemarketers' representations. When the item is supposed to be a tangible "gift" or "prize" of substantial value, as in charity schemes or prize-promotion schemes, fraudulent

---

"What Is Telemarketing Fraud?" U.S. Department of Justice, Criminal Division, http://www.usdoj.gov/criminal/fraud/telemarketing/whatis.htm. September 25, 1998. Downloaded March 10, 2002. Although this document is dated 1998, readers will find the information contained within it useful.

telemarketers will instead provide what they term a "gimme gift." The "diamond watch" that the consumer thought would be worth many hundreds or thousands of dollars, for example, proves to be an inexpensively produced watch with a small diamond chip, for which the fraudulent telemarketer may have paid only $30 to $60.

(2) *To obtain immediate payment before the victim can inspect the item of value they expect to receive.* Regardless of what good or service a fraudulent telemarketer says he is offering—investment items, magazine subscriptions, or office supplies, for example—a fraudulent telemarketer will always insist on advance payment by the consumer before the consumer receives that good or service. If consumers were to receive the promised goods or services before payment, and realized that the good or service was of little or no value, most of them would likely cancel the transaction and refuse payment.

Fraudulent telemarketers therefore routinely make false and misleading representations to the effect that the consumer must act immediately if he or she is to receive the promised good or service. These representations may suggest that the opportunity being offered is of limited quantity or duration, or that there are others also seeking that opportunity. In addition, fraudulent telemarketers usually persuade the victims to send their money by some means of expedited delivery that allows the telemarketers to receive the victims' payments as quickly as possible. For victims who have checks or money orders, the telemarketers use nationally advertised courier delivery services, which will deliver victims' checks by the next business day. For victims who have credit cards, the telemarketers obtain merchant accounts at financial institutions, so that the credit-card number can be processed immediately.

(3) *To create an aura of legitimacy about their operations, by trying to resemble legitimate telemarketing operations, legitimate businesses, or legitimate government agencies.* Magazine-subscription schemes, for example, often tell consumers, "We're just like" a nationally publicized magazine-distribution organization, and in some cases have simply lied to consumers by stating that they are the nationally publicized organization. Telemarketers in "rip-and-tear" schemes or "recovery-room schemes" often falsely impersonate federal agents or other government officials to lend greater credibility to their demands for money.

Another factor that distinguishes fraudulent from legitimate telemarketing operations is "reloading." "Reloading" is a term that refers to the fraudulent telemarketer's practice of recontacting victims, after their initial transactions with the telemarketer, and soliciting them for additional payments. In prize-promotion schemes, for example, victims are often told that they are now eligible for even higher levels and values of prizes, for which they must pay additional (nonexistent) "fees" or "taxes." Because "reload" transactions typically demand increasingly substantial amounts of money from victims, they provide fraudulent telemarketers with their most substantial profits, while causing consumers increasingly large losses that they will never recoup voluntarily from the fraudulent telemarketers.

A third factor that distinguishes fraudulent from legitimate telemarketing operations is the fraudulent telemarketer's general reluctance to contact prospective victims who reside in the state where the telemarketing operation conducts its business. Fraudulent telemarketers recognize that if they contact victims located outside their state, any victims who later realize that they may have been defrauded are likely to be uncertain about which law enforcement agency they should contact with complaints, and less likely to travel directly to the telemarketing operation and confront the telemarketers about their losses.

Although many consumers apparently find it difficult to believe that there are people who will contact them on the telephone and lie and misrepresent facts in order to get their money, the reality is that at any given time, there are at least several hundred fraudulent telemarketing operations—some of them employing as many as several dozen people—in North America that routinely seek to defraud consumers in the United States and Canada. Moreover, these schemes generally do not choose their victims at random. Fraudulent telemarketers routinely buy "leads"—that is, listings of names, addresses, and phone numbers of persons who have been defrauded in previous telemarketing schemes (and typically the amount of their last transaction with a fraudulent telemarketer)—from each other and from "lead brokers," companies that engage exclusively in buying and selling fraudulent telemarketers' leads. Although leads are relatively costly to the fraudulent telemarketer—as much as $10 or even $100 per lead in some cases—they also indicate to the fraudulent telemarketer which consumers are most likely to be persuaded to send substantial amounts of money that will far exceed the cost of the leads.

Firms giving references may provide the names of "touts" or "singers." "Touts" and "singers" are people who praise the telemarketer's services, but who actually are part of the scheme. Telemarketers also sometimes give as a reference an organization with a name similar to the "Better Business Bureau" (BBB), but which in reality has nothing to do with a legitimate local BBB.

Chapter 25

What Kinds of Telemarketing Schemes Are out There?

In one sense, the nature and content of a telemarketing scheme to defraud is limited only by the ingenuity and skill of the scheme's organizers. Fraudulent telemarketers who observe that certain types of business, or certain business trends, are widely publicized in news media will incorporate references to such reports in their solicitations of victims. In practice, most fraudulent telemarketers operate one or more of the following types of schemes: charity schemes; credit-card, credit-repair, and loan schemes; cross-border schemes; Internet-related schemes; investment schemes; lottery schemes; office-supply schemes; "prize-promotion" schemes; "recovery-room" schemes; and "rip-and-tear" schemes.

Charity Schemes

Many people have a laudable desire to help those who are less fortunate by making donations to charitable causes. Fraudulent telemarketers have often exploited that desire, by devising schemes that purport to raise money for worthy causes. At various times, fraudulent telemarketers have falsely claimed, for example, that they were soliciting funds for victims of the Oklahoma City bombing or Mississippi

---

"What Kinds Of Telemarketing Schemes Are Out There?," U.S. Department of Justice, Criminal Division, http://www.usdoj.gov/criminal/fraud/telemarketing/schemes.htm. September 25, 1998. Downloaded March 10, 2002. Although this document is dated 1998, readers will find the information contained within it useful.

Valley flooding or for antidrug programs. In one type of charity scheme, known as "badge fraud," fraudulent telemarketers purport to be soliciting funds to support police-or fire department-related causes.

Some of these schemes simply lie outright to would-be donors and give none of their proceeds to charity. Other schemes, to help maintain their aura of legitimacy, will donate a modest amount of their proceeds—typically no more than 10 percent—to a charitable cause so that they can show some evidence of their "legitimacy" to law-enforcement or regulatory authorities who may have received complaints about the schemes. In one scheme that resulted in successful federal criminal prosecutions, the schemes' organizers tried to enhance their appearance of legitimacy by sending the would-be donors "gimme gifts" and mounted plaques that thanked them for their contributions to a particular "foundation" (which was in fact an organization, run by former fraudulent telemarketers, that did nothing for people in need).

Credit-Card, Credit-Repair, and Loan Schemes

Certain telemarketing schemes are devised to victimize people who have bad credit, or whose income levels may be too low to allow them to amass substantial credit. In many credit-card telemarketing schemes, telemarketers contact prospective victims and represent that they can obtain credit cards even if they have poor credit histories. The victim who pays the fee demanded by the telemarketer usually receives no card, or receives only a credit-card application or some cheaply printed brochures or flyers that discuss credit cards. In a variant of these schemes, the credit card that some consumers are given, after paying the fee to the telemarketer, requires the consumer to pay a company located outside the United States $200 or $300 and limits the consumer's charges on that card to an amount no greater than the amount of money the consumer has paid to the offshore company.

In advance-fee loan schemes, persons with bad credit are promised a loan in return for a fee paid in advance. Victims who pay the fees for the guaranteed loans have their names referred to a "turn-down room"—that is, an operation, affiliated with the telemarketer, whose sole function is to notify the victims at a later date that their loan applications have been rejected.

Another way that fraudulent telemarketers try to take advantage of people with poor credit histories is by promising to "repair" their credit. Firms that promise that they can remove bankruptcies, judgments, liens, foreclosures and other items from a consumer's credit report, irrespective of the age or accuracy of the information, are misstating the facts. Judgments, paid tax liens, accounts referred for collection or charged off, and records of arrest, indictment, or conviction, may remain on a consumer's report for seven years, while bankruptcies may remain on the report for ten years. Credit reporting agencies are obligated by law to correct mistakes on consumers' reports. People with real errors on their reports can deal directly with the reporting agencies, or obtain assistance from the Federal Trade Commission with their problems. Credit repair agencies cannot generally succeed in eliminating accurate negative information from credit reports, despite the contrary claims of some firms.

Cross-Border Schemes

Cross-border telemarketing schemes consist of telemarketing schemes—usually advance-fee loan schemes, investment schemes, lottery schemes, and prize-promotion schemes—where the scheme's operators conduct their telemarketing activities in one country and solicit victims in another country. Cross-border telemarketing schemes are appealing to some telemarketers because it further enhances the difficulties for consumers in reporting complaints and for law-enforcement agencies in investigating and prosecuting these schemes. To obtain certain kinds of evidence that they can use in legal proceedings, and to have criminal suspects extradited from foreign countries, law enforcement and regulatory authorities in the United States must use legal procedures that have been established by bilateral treaties with those countries. In some cases, those procedures provide opportunities for the criminal suspects to create substantial delays in legal proceedings by challenging the U.S. authorities' right to the evidence or the extradition of the suspect. A few fraudulent telemarketers have admitted to law enforcement authorities that they believe these delays can last long enough for victim-witnesses to die before they have an opportunity to testify in legal proceedings and receive some restitution of their losses.

Internet-Related Schemes

Many of the schemes to defraud that have been conducted exclusively by telemarketing are now being conducted partly or exclusively by the Internet. Consumers who have Internet access now regularly receive "spammed" e-mails—that is, unsolicited e-mails that purport to offer business or investment opportunities or opportunities to purchase computer-related or other goods or services. Much of the "spamming" invites consumers to telephone a particular telephone number, to visit a particular webpage, or to mail funds to a particular address.

Many of these spammed e-mails, however, are sent by fraudulent schemes. In the past year, for example, federal and state authorities have actively pursued fraudulent schemes advertising on the Internet that included promotions for charities, investments such as pyramid schemes and Ponzi schemes, degrees from a fictitious educational institution, illegal raffles, and opportunities to sell coupon certificate booklets or to earn money at home by clipping coupons. According to the Internet Fraud Watch, a project of the National Consumers League, fraud reports to the Internet Fraud Watch increased substantially from an average of 32 per month in 1996 to nearly 100 per month in 1997. The Internet Fraud Watch also reported that for the period January–July 1997, the top ten categories of Internet-related fraud complaints were:

1. Internet and online services that were misrepresented or never delivered;
2. General merchandise that was never delivered or not as advertised;
3. Auctions of items that were never delivered or whose value was inflated;
4. Pyramid- and multilevel marketing schemes in which profits are made from fees to join the scheme rather than sales of actual items;
5. Business opportunities that were substantially less profitable than advertised;
6. Work-at-home schemes that sold materials to consumers with false promises to pay for the work performed;
7. Prizes and sweepstakes schemes in which prizes were never awarded;
8. Credit-card offers in which consumers never received the promised cards;

9. Sales of self-help manuals that were misrepresented or never delivered; and

10. Magazine subscriptions that were never delivered or for which the scheme's affiliation with legitimate publishers was misrepresented.

Investment and Business-Opportunity Schemes

Since the 1970s, many fraudulent telemarketers have offered a wide variety of spurious investment opportunities to would-be investors. The nature of the purported investments varies largely with what the telemarketers perceive, from widely publicized reports on business matters, to be the trends that less experienced or less sophisticated investors are most likely to consider attractive for high-profit investment. In the 1980s, for example, many telemarketing schemes offered "investments" in rare coins, precious metals, and so-called "strategic metals." In the 1990s, many telemarketing schemes turned to offering "investments" in items as diversified as "investment-grade" gemstones, ostrich farms, and telecommunications technology such as wireless cable systems for television broadcasting.

Where the investment item that the victim purchases is small enough to be held, such as gemstones or rare coins, fraudulent telemarketers often seal the items in a plastic container and ship the sealed container to the victim. The telemarketer also warns the victim not to open the container, as that would void the telemarketer's guarantee on those items. The true purpose of sealing the items in the container is to dissuade the victim from having the items independently appraised, as a genuine appraisal would inform the victim that the real market value of the items is far below what the telemarketer had represented. Where the investment items are too large to be provided directly to the victim, such as oil and gas wells, bars of precious metals, or wireless cable systems, the telemarketers often lie to the victims about the fact that the investment item in fact does not exist or is not yet developed enough to turn a profit.

In a number of investment telemarketing schemes, the operators "reload" their victims by representing that they have prospective buyers who are interested in the investment that the victim has made with the telemarketer, and that the victim must buy more of the same item, or a particular type of item (such as more gemstones), to "round out" their investment portfolio. The victim is later told that the prospective buyer decided not to purchase the victim's investment, leaving the victim with still greater losses and no realistic prospect of recouping those losses.

Business-opportunity schemes are a variant of investment schemes. These prey on people who would like to be in business on their own. Telemarketers frequently promise to set consumers up in business using a formula or method that is supposed to be virtually certain to guarantee success. Many of these businesses involve installing vending machines, toy carousels, computer or video games, pay telephones, and the like. Purchasing such business opportunities can cost many thousands of dollars, but often leaves the consumer with a basement full of products and machines he or she was unable to sell.

A common method of operation is for the telemarketer to take out ads in local newspapers, with an 800 number for interested consumers to call. The ad may or may not make exaggerated earnings claims, like "Earn $100,000 in your spare time." Consumers who call the 800 number often hear more about potential earnings, and lots of promises regarding the success of the business. All too often, unfortunately, there is nothing to back up these claims.

A legitimate seller of franchises will provide consumers with detailed information about others who have purchased the franchise, the basis for earnings claims, and other similar information. The list of franchisees provided to the consumer should include at least the 10 franchisees closest to the consumer, if there are 10 franchisees. Consumers should talk to other franchisees to learn about the business. Generally speaking, promises of a lot of money for little effort only describe what the telemarketer will get, not what a consumer will end up with.

Lottery Schemes

Fraudulent telemarketers have often used offers of "investing" in foreign lottery tickets or chances as a vehicle to defraud consumers. Although federal law forbids the importation, interstate transportation, and foreign transportation of lottery tickets or chances, many fraudulent telemarketers routinely contact people in the United States, through mailings, advertisements, and telephone calls, to solicit their involvement in foreign lotteries such as the "Australian Lottery" or "El Gordo." Victims often begin by paying as little as $5 or $10 for lottery chances. Many of those who do are later contacted by telemarketers who hold themselves out as "experts" in "investing" in lottery chances, and who solicit the victims for larger and larger amounts of money. Law enforcement authorities in the United States and Canada are aware of many instances in which victims have sent thousands, tens of thousands, and even hundreds of thousands of dollars to lottery telemarketers, after hearing repeated promises or guarantees of vast returns.

In reality, the telemarketers invest little or none of the victims' money in the lottery tickets or chances, and keep the money for themselves after paying salaries and other expenses of their fraudulent business. In some instances, where victims had been given certain lottery numbers by the telemarketers, and later learned independently that their numbers had won a real lottery, the victims were told that their winnings had been "invested" in other lottery tickets rather than being paid to them directly.

Magazine-Promotion Schemes

A number of fraudulent telemarketing schemes in recent years have turned to offering magazine subscriptions, in part because of the apparent popularity of several nationally known magazine-promotion businesses that send mailings throughout the United States that advertise their multimillion-dollar prize contests. In a typical fraudulent magazine-promotion scheme, a telemarketer contacts a prospective victim and tells the victim that he or she has won a highly valuable prize or is to be given a highly valuable "gift," and falsely implies that the victim must purchase multiple magazine subscriptions to receive the award or gift. The victim is told the total price for the package of subscriptions, which can range from several hundred to several thousand dollars, but is not told which magazines he or she is to receive when the victim agrees to send money to the telemarketer. Instead, the telemarketer, after receiving the victim's money, sends the victim lists of magazines on which the victim can check off the magazines that he or she wants to order and which the victim then mails back to the telemarketer.

Although magazine-promotion schemes typically claim that they are "just like" the nationally advertised magazine-promotion businesses, consumers should note that there are three critical differences between fraudulent and legitimate magazine-distribution companies:

Advance Disclosure of Prices

In the promotional materials that they mail to consumers, legitimate magazine-promotion companies typically disclose the newsstand price, the discounted price, and the savings to the consumer in advance of any magazine purchase by the consumer. In contrast, fraudulent magazine-subscription schemes do not disclose the newsstand price or the true price per issue that they are charging their victims; to disclose those two prices to the prospective victim would immediately inform the victim that the price is far above even the newsstand price or, in some cases, that the magazine being offered is free and has no subscription price.

Advance Disclosure of Magazine Titles

Legitimate magazine-promotion companies clearly inform consumers of the magazine titles that they can order, before the consumer has to send any money. In contrast, fraudulent magazine-promotion schemes do not disclose these titles in advance; to do so would reveal that many of the most widely circulated and popular magazines are not available through the scheme.

Provision of All Magazines Promised

Legitimate magazine-subscription companies provide all of the magazines that a customer has ordered, and for the full duration of the subscription. In contrast, when they solicit prospective victims, fraudulent telemarketers—sometimes telling prospective victims that no one can read dozens of magazines each month—represent to victims that their company will donate the magazines to a local hospital or hospice, so that less fortunate people can benefit from the victims' purchase. In reality, fraudulent telemarketers frequently choose not to subscribe to any magazines at all in the victim's name, or submit the victim's subscription for one year rather than two and pocket the remainder of the victim's money.

Office-Supply Schemes

One type of fraudulent telemarketing scheme that is directed specifically at the business community is the office-supply or "toner" scheme. In a typical toner scheme, an employee of the telemarketer first contacts a business under false pretenses to learn the make and model number of the copier used at that business. Within a day or two, a telemarketer then recontacts the business, this time falsely representing himself to be a representative of the copier company whose copier is being used at that business. The telemarketer falsely states that his company is about to increase the price of their toner, but that the company is prepared to offer the business a shipment of toner at the then-current price before the price increase goes into effect. If the business employee who answers the telemarketer's call agrees to have the toner shipped, the business thereafter receives an invoice for the toner without the toner, or receives a shipment of toner with the invoice.

In the latter case, the toner is not only priced far higher per unit than the real copier company charges for its product, but also is mislabeled to leave the impression that it can be used in the type of copier in use at the business. In fact, the toner is usually what is known as "graymarket" toner—that is, toner that does not meet the quality specifications and requirements of legitimate copier companies. Businesses who use the graymarket toner often find that it clogs their copiers, requiring them to call their copier company for service. In many cases, businesses do not discover the true extent of the fraud until the copier company's service people repair the copiers and inform the businesses that the toner is not their company's toner and that there was no planned price increase.

Prize-Promotion Schemes

Over the years, law-enforcement and regulatory authorities have observed at least three varieties of telemarketing schemes that purport to offer prizes, awards, or gifts to consumers if the consumers buy certain goods or services or make supposedly charitable donations:

(1) The oldest of these schemes, which still is carried out in some areas of both the United States and Canada, involves a simply false statement to a prospective victim that he or she has won a particular highly valuable item, such as a car. Because the explicit promise of a valuable item that is never delivered provides clear evidence of fraudulent intent, participants in a number of these schemes were successfully prosecuted in the 1980s and 1990s.

(2) More recently, beginning in the late 1980s and early 1990s, some fraudulent telemarketers began conducting so-called "one-in-five" schemes. In a "one-in-five" scheme, the telemarketer contacts a prospective victim and represents that the victim has won one of five (sometimes four or six) valuable prizes. The telemarketer then lists the prizes for the victim. All but one of the promised prizes would in fact have substantial value if awarded (such as a nationally known U.S.-manufacture car, a $3,000 cashier's check, and a $5,000 cashier's check); the remaining prize, however, was a "gimme gift" such as inexpensive "his and hers diamond watches," plastic dolphin statues, and gold rings.

Many telemarketers, in reading the award list to victims, deliberately included the "gimme gift" between two items of much higher real value, to make the victim believe that the "gimme gift's" value was somewhere between the value of the other two items. Although most "one-in-five" schemes invariably gave away only the "gimme gift," the telemarketers typically told victims who asked which prize they would be receiving that they could not give out that information because it would be "collusion" or "bribery." These explanations, while nonsensical, were developed to avoid having to tell prospective victims clearly false information about the prizes' value and the victims' chances of winning the far more valuable items.

(3) A variation on the "one-in-five" scheme is the so-called "mystery pitch" or "integrity pitch" (the latter term being more favored by fraudulent telemarketers). In a "mystery pitch," the telemarketer would tell a prospective victim that he or she had won something of substantial value, but refuse to tell the victim what the prize was because of professed concerns about "collusion" or "bribery." Some telemarketers apparently adopted the "mystery pitch" after seeing that other telemarketers were being successfully prosecuted in "one-in-five" schemes where the prizes were specifically listed for the victims.

In each of these variations of "prize-promotion" schemes, the telemarketers routinely "reload" victims by telling them that they have now qualified for higher levels of prizes, but that they must accordingly pay still more money to the telemarketers to cover the "fees" or "taxes" that the telemarketers claim are due on the prizes.

"Recovery-Room" Schemes

So-called "recovery-room" telemarketing schemes are schemes that often are extensions of other telemarketing schemes. As one or more telemarketing schemes gradually deprive victims of most of their funds, the victims become increasingly desperate to try to recover a portion of their losses, and increasingly concerned about the embarrassment that they would feel if they had to report the true extent of their losses to law enforcement. A telemarketer for a "recovery room" contacts the victim, and invariably claims some affiliation with a government organization or agency that is in a position to help telemarketing victims recover some of their past losses. Some telemarketers have falsely impersonated FBI, IRS, and Customs agents or attorneys in law firms, while others use more nebulous language to suggest a connection to a government agency (such as "I'm working with the court [or the district attorney]") in a particular city.

Because the telemarketers may have worked for the scheme that had previously defrauded the victims, they are well-equipped to disclose information about the amounts of the victims' past losses. This in turn gives the telemarketers additional credibility with the victims, who apparently believe that only a legitimate law-enforcement or other government agency could have access to such information. The telemarketers, posing as government officials, then tell victims that they must send a fee to them so that their money can be released by "the court" or otherwise delivered to the victims. This allows the telemarketer to deprive victims of even more money, while simultaneously encouraging the victims to believe that something is being done to protect their interests and causing the victims to postpone the filing of complaints with real law-enforcement or regulatory authorities.

"Rip-and-Tear" Schemes

As law-enforcement and regulatory authorities have become more vigorous in prosecuting fraudulent telemarketing, some fraudulent telemarketers have increasingly engaged in what are known as "rip-and-tear" schemes. What distinguishes a "rip-and-tear" scheme from other telemarketing schemes is simply the methods that the fraudulent telemarketers use to minimize the risks of detection and apprehension of the authorities. Instead of conducting their telemarketing from a single, fixed place of business, "rip-and-tear" telemarketers conduct their calls from various places, such as pay telephones, residences, and hotel rooms. Their contacts with prospective victims—who usually are repeat victims of past telemarketing schemes—involve explicit promises that the victims have won a valuable prize or are entitled to receive a portion of their past losses.

"Rip-and-tear" schemes often insist that the victims send the required "fees" to commercial mailbox facilities or by electronic wire transfer services, which create far less substantial paper trails than checks or credit cards and which allows the telemarketers to receive the victims' payments in cash. To create further difficulties for law enforcement, "rip-and-tear" telemarketers often hire persons to act as couriers to pick up the payments from the mailbox drop or wire transfer office; if law-enforcement agents pick up the payments and arrest the couriers, the organizers of the scheme are not tied directly to the delivery of the victims' funds. In some instances, law-enforcement authorities have reported that telemarketers appeared to be conducting counter surveillance—that is, watching to see whether they were under surveillance at pickup points, and engaging in evasive behavior.

Chapter 26

Unwanted Telephone Marketing Calls

Has your evening quiet time been disrupted or your dinner been interrupted by a call from a telemarketer? If so, you're not alone. Consumers are increasingly complaining to the Federal Communications Commission (FCC) about unwanted and uninvited calls to their homes from advertisers and telemarketers.

The Telephone Consumer Protection Act (TCPA) of 1991 was created in response to consumer concerns about the growing number of unsolicited telephone marketing calls to their homes and the increasing use of automated and prerecorded messages. The FCC has rules to aid consumers who wish to limit these uninvited calls.

Telephone Solicitations

A "telephone solicitation" is a telephone call that acts as an advertisement. Even if you have an unlisted, non-listed, or non-published telephone number, you may still receive unsolicited telephone calls. In some cases unlisted or non-listed numbers can be obtained from a directory assistance operator. They, along with non-published numbers, may be sold to other organizations or people with whom you have done business. Some sales organizations call all numbers in numerical order for a neighborhood or area.

"Unwanted Telephone Marketing Calls," Federal Communications Commission (FCC), http://www.fcc.gov/cgb/consumerfacts/tcpa.html. Updated February 13, 2002.

The FCC's rules prohibit telephone solicitation calls to your home before 8 am or after 9 pm. Anyone making a telephone solicitation call to your home must provide his or her name, the name of the person or entity on whose behalf the call is being made, and a telephone number or address at which that person or entity may be contacted.

The term "telephone solicitation" does not include calls or messages placed with the receiver's prior consent, regarding a tax-exempt nonprofit organization, or from a person or organization with whom the receiver has an established business relationship. An established business relationship exists if you have made an inquiry, application, purchase, or transaction regarding products or services offered by the person or entity involved. Generally, you may put an end to that relationship by telling the person or entity not to place any more solicitation calls to your home.

Automatic Telephone Dialing Systems and Artificial or Prerecorded Voice Calls

Automatic telephone dialing systems, also known as "autodialers," also generate a lot of consumer complaints.

Autodialers produce, store and dial telephone numbers using a random or sequential number generator. Autodialers are usually used to place artificial (computerized) or prerecorded voice calls. Except for emergency calls or calls made with the prior express consent of the person being called, autodialers and any artificial or prerecorded voice messages may not be used to contact numbers assigned to:

- any emergency telephone line;
- the telephone line of any guest or patient room at a hospital, health care facility, home for the elderly, or similar establishment;
- a paging service, cellular telephone service, or other radio common carrier service, if the person being called would be charged for the call; or
- any other service for which the person being called would be charged for the call.

Calls using artificial or prerecorded voice messages—including those that do not use autodialers—may not be made to residential telephone numbers except in the following cases:

- emergency calls needed to ensure the consumer's health and safety;
- calls for which you have given prior consent;
- non-commercial calls;
- calls which don't include any unsolicited advertisements;
- calls by, or on behalf of, tax-exempt non-profit organizations; or
- calls from entities with which you have an established business relationship.

Calls using autodialers or artificial or prerecorded voice messages may be placed to businesses, although the FCC's rules prohibit the use of autodialers in a way that ties up two or more lines of a multi-line business at the same time.

If an autodialer is used to deliver an artificial or prerecorded voice message, that message must state, at the beginning, the identity of the business, individual, or other entity initiating the call. During or after the message, the caller must give the telephone number (other than that of the autodialer or prerecorded message player that placed the call) or address of the business, other entity, or individual that made the call. It may not be a 900 number or any other number for which charges exceed local or long distance transmission charges.

Autodialers that deliver a recorded message must release the called party's telephone line within 5 seconds of the time that the calling system receives notification that the called party's line has hung up. In certain areas there might be a delay before you can get a dial tone again. Your local telephone company can tell you if there is a delay in your area.

How to Reduce the Number of Telephone Solicitation Calls Placed to Your Home

The FCC requires a person or entity placing live telephone solicitations to your home to maintain a record of your request not to receive future telephone solicitations from that person or entity. A record of your do-not-call request must be maintained for ten years. This request should also stop calls from affiliated entities if you would reasonably expect them to be included, given the identification of the caller and the product being advertised. Each time you receive a call from a different person or entity, though, you must request that that person or entity not call you again. Tax-exempt non-profit organizations are not required to keep do-not-call lists.

When you receive telephone solicitation calls, clearly state that you want to be added to the caller's do-not-call list. You may want to keep a list of those places that you have asked not to call you.

The Direct Marketing Association (DMA) sponsors the Telephone Preference Service (TPS) which maintains a do-no-call list. DMA members are required to use this list. Once you register, your name stays on file for 5 years. [See the "Additional Help and Information" section of this Sourcebook for contact information.]

Registering your number with the TPS should prevent sales calls from all companies that belong to the DMA. Questions about the DMA's registration program should be sent to the DMA at this address. While registering with the DMA should reduce the number of unsolicited calls you receive at your home, it probably will not stop unwanted calls altogether.

Finally, many states now have statewide "no-call" lists for residents in that state. Contact your state's consumer protection office or the public utilities commission (PUC) to see if your state has such a list.

Actions You Can Take against Telephone Solicitation Calls Made in Violation of the Rules

You have recourse against entities or persons who continue to call you after you have requested to be placed on a "do not call" list. Some states permit you to file law suits against the violators; you may be awarded $500 in damages or actual monetary loss, whichever is greater. This amount may be tripled if you are able to show that the caller willfully and knowingly violated do-not-call requirements.

States themselves may initiate a civil suit in federal district court against any person or entity that engages in a pattern or practice of violations of the TCPA or FCC rules. If you have questions for your state regarding unsolicited telephone marketing, you may contact your local or state consumer protection office or your state Attorney General's office. These numbers should be listed in the government section of your telephone directory, or you can obtain them by calling directory assistance.

What the FCC Can Do to Help

While the FCC may not award monetary or other damages, it can give citations or fines to those violating the TCPA or other FCC rules regarding unsolicited telephone marketing calls. [See the "Additional Help and Information" section of this *Sourcebook* for further information on filing a complaint with the FCC.]

Chapter 27

What Can I Do about Telemarketing Fraud?

Because telemarketing, both legitimate and fraudulent, occurs on a daily basis with people throughout the United States and Canada, consumers need to think about how they ought to respond to the telemarketing calls they receive, or what they should do if they realize that they or someone they know may have been defrauded by a telemarketer. Some of the questions that consumers are most likely to ask are listed below.

How Can I Find out Whether a Telemarketer Who Calls Me Is Legitimate or Not?

The Department of Justice and the FBI do not maintain any list of "legitimate" or "fraudulent" telemarketers that consumers can consult. Such a list would be impossible for the Department to maintain in a current and accurate form, as many legitimate businesses and fraudulent schemes are beginning and ending operations throughout North America at any given time. As a general proposition, consumers should always take the time to investigate carefully any request or demand for money that any telemarketer makes, and should never be rushed or forced into making a hasty decision.

If, for example, someone on the telephone claims to be an FBI agent or Department of Justice attorney and insists that you send money,

---

"What Can I Do About Telemarketing Fraud?," U.S. Department of Justice (DOJ), http://www.usdoj.gov/criminal/fraud/telemarketing/action.htm, last updated February 10, 2000.

you should know that no federal law-enforcement agent or Department of Justice attorney is ever authorized or permitted to ask you for any "fees" or "taxes" if you have lost money to fraudulent telemarketers in the past. It is part of the Department's and the FBI's day-to-day responsibility, in investigating and prosecuting fraud, to try to recover victims' losses, but the Department and the FBI carry out those responsibilities with the funding that Congress provides through the appropriations process. You should therefore feel free to take the name and telephone number of the person who claims to be an agent, tell him that you will call him back after verifying his statement, and then contact the nearest FBI office or FBI headquarters (202-324-3000) or the Department of Justice (202-353-1555) and explain why you are calling.

One organization that consumers can contact to ask about telemarketing calls they receive is the National Fraud Information Center (NFIC). The NFIC maintains an "800" toll-free number that consumers can call if they have questions or want to report possible instances of telemarketing fraud. The NFIC number is 800-876-7060. In addition, the Federal Trade Commission's Consumer Response Center maintains numbers that consumers can call with questions or complaints about all types of consumer fraud, including telemarketing fraud. The Consumer Response Center numbers are 202-382-4357 and 202-326-3128.

How Can I Stop Getting Telemarketing Calls?

If you want to stop receiving calls from a particular telemarketer, whether or not you think the telemarketer is legitimate or fraudulent, federal law allows you to ask that telemarketer to put your name on his company's "do-not-call" list. The Federal Trade Commission's Telemarketing Sales Rule requires the telemarketer to comply with your request. A telemarketer who fails to remove your name from any list that it uses to contact consumers by telephone, and who continues to contact you even after receiving your request, may be subject to civil litigation by the FTC or by a state attorney general to enforce the Telemarketing Sales Rule, and could be given civil money penalties of as much as $11,000 per violation.

If you want to stop receiving telemarketing calls in general, you should contact the Direct Marketing Association (DMA), which represents many businesses that engage in telemarketing and other forms of direct sales, and request that your name be removed from lists that their members use. While many fraudulent telemarketers will not respect such a request, the DMA will comply with your request. The DMA's telephone numbers are 212-768-7277 [New York] and 202-955-5030 [Washington, D.C.].

What Should I Do If I Think Someone I Know Is Being Defrauded by a Telemarketer?

If you think that you may have been defrauded by a telemarketer, you should report your information as soon as possible. Some people who have lost substantial amounts of money have refrained from reporting their losses, in part because they fear embarrassment and critical reaction from their family, friends, or the people to whom they would be reporting the fraud. Please note that fraudulent telemarketers count on that reaction: the longer that people refrain from reporting a fraudulent telemarketer's activities, the longer that telemarketer can continue to operate and the more people the telemarketer will try to defraud.

If you contact any of the following organizations with reports or suspicions about possible telemarketing fraud, you will be treated with respect and your report will be taken promptly and made available promptly to a number of federal, state, and local law-enforcement and regulatory agencies that may be in a position to act on your report:

- National Fraud Information Center (NFIC)—Telephone: 800-876-7060
- Federal Trade Commission Consumer Response Center—Telephone: 202-382-4357 or 202-326-3128.

If you know that someone in your family, such as a sibling or a parent, appears to have been defrauded by a telemarketer, you should avoid confronting that person directly and stating that he or she has been "duped" or "swindled." Particularly in cases where an older parent may be the victim of telemarketing fraud, the victim is likely to fear that others will try to take away his or her independence or to take over their affairs, and may become highly resistant to your efforts to address the situation. Unfortunately, a number of adult children of older parents who were victimized have reported that their efforts to try to confront the problem with their older parents resulted only in creating severe strains on, and even apparently irreparable damage to, their relationships with the victims. Some fraudulent telemarketers have been known to persuade victims that they are more concerned about the victims' welfare than the victims' own family, whom the telemarketers try to portray as greedy.

A better tactic is to approach the matter indirectly and exercise patience in talking with the person about your concerns. It should not offend the person's dignity if you mention that people have to be careful about telemarketing scam artists, and suggest that it is wise to consult other sources of information about investment opportunities (or whatever the telemarketing scheme appears to be offering) before trusting your money to someone else. Before they have come to terms with the fact that they have been defrauded, telemarketing fraud victims are often reluctant to talk with others about their dealings with the telemarketer, or to admit how many transactions they have done with the telemarketer or how much money they have actually lost. Patience is therefore essential in establishing and maintaining a dialogue with the possible victim, and in trying to learn as much information as possible that could help in deciding whether the situation should be reported to the authorities. Again, statements that may sound judgmental or critical to telemarketing fraud victims are likely to stiffen the victims' resistance to talking about their dealings with the telemarketers, and to make them suspicious of your motives in raising the subject.

How Can I Report Possible Telemarketing Fraud?

Consumers anywhere in the United States or Canada who want to telephone in their reports of suspected telemarketing fraud should call the National Fraud Information Center (NFIC), toll-free, at 800-876-7060, or the Federal Trade Commission's Consumer Response Center at 202-382-4357 and 202-326-3128. Consumers who want to file their reports or complaints on-line can contact the NFIC [www.fraud.org] or the U.S. Consumer Gateway [www.consumer.gov], which gives consumers access to several federal agencies. In addition, consumers within the United States can contact the nearest office of the FBI or FBI headquarters (202-324-3000).

Where Can I Learn More about Telemarketing Fraud?

Many federal, state, and local government agencies and private-sector organizations are compiling and issuing information about telemarketing fraud on a regular basis, including information available through their webpages. If you would like to visit some of these webpages, please feel free to use the following list, which will give you a starting point for further information on telemarketing fraud and related subjects [See the Additional Help and Information section of this *Sourcebook* for contact information for the following organizations]:

U.S. Agencies and Organizations

- American Association of Retired Persons
- Better Business Bureau
- Commodity Futures Trading Commission
- Federal Bureau of Investigation
- Federal Trade Commission
- Internet Fraud Watch
- National Association of Attorneys General
- National Consumers League
- National Fraud Information Center
- North American Securities Administrators Association
- U.S. Customs Service
- U.S. Postal Inspection Service
- U.S. Securities and Exchange Commission
- U.S. Sentencing Commission

Chapter 28

The Federal Trade Commission and the National "Do Not Call" Registry

Current Consumer Protections

Under the Telemarketing Sales Rule, it is illegal for a telemarketer to call you once you've asked them not to. If they call you again anyway, report them to your state Attorney General.

The TSR also:

- Restricts calling times to the hours between 8 a.m. and 9 p.m.
- Requires telemarketers to tell you it's a sales call and who's doing the selling before they make their pitch.
- Prohibits telemarketers from lying or misrepresenting any information.

The National "Do Not Call" Registry

Under the FTC's change to the TSR, a consumer will be able to call a toll-free number to place their phone number on a national "do not call" registry. Once your number is on the registry, it will be illegal for a telemarketer to call it.

This chapter contains text from "Consumer Update: The FTC's Proposal to Create a National 'Do Not Call' Registry," Federal Trade Commission (FTC), http://www.ftc.gov/bcp/conline/pubs/alerts/dncalrt.htm. Dated 2002. This chapter also contains text from "Facts for Consumers: You Make the Call: The FTC's New Telemarketing Sales Rule," Federal Trade Commission (FTC), http://www.ftc.gov/bcp/conline/pubs/tmarkg/donotcall.htm, December 18, 2002.

Telemarketers will be required to "scrub" their lists, removing the numbers of all consumers who placed themselves on the national "do not call" registry. The FTC will manage the registry.

Placing your number on the "do not call" registry will stop most, but not all, telemarketing calls. Certain businesses are exempt from the TSR, and thus can still call you even if you place your number on the registry—including common carriers (such as long-distance phone companies and airlines) and insurance companies operating under state regulations. Also, according to the proposal, an individual company would be allowed to call you, even if you placed your number on the registry, as long as you gave the company your express verifiable authorization to do so—for example, by giving them written permission to call you.

How to Reduce Unwanted Telemarketing Calls Now

When the FTC begins to implement a national "do not call" registry, it will be months before it takes effect. However, there are steps you can take right now to reduce the number of unwanted telemarketing calls that you receive.

1. Ask companies that call you to put you on their company-specific "do not call" list. Existing regulations by both the FTC and the FCC prohibit a telemarketer from calling you after you have asked them to stop calling you.

2. Register with a state "do not call" list. Many states offer "do not call" lists for residents of that state. Rules for how to put your name and number on the list and which telemarketers are covered vary. [See the "Additional Help and Information" section of this *Sourcebook* for the FTC's contact information. The FTC maintains a list of current state contact information.] Other states are creating "do not call" lists, while still others encourage consumers to use the Direct Marketing Association's (DMA) Telephone Preference Service.

3. Register with the Direct Marketing Association's Telephone Preference Service. The DMA's Telephone Preference Service (TPS) allows consumers to opt out of receiving telemarketing calls from many national companies for five years. When you register with the DMA's TPS, your name is put on a "delete" file that is updated four times a year and made available to telephone marketers. You should notice a decrease in the

number of telemarketing calls you receive about two or three months after your name is entered into the quarterly file. Remember that your registration will not stop calls from organizations that are not registered with the DMA's Mail and Telephone Preference Services. To delete your name from many telemarketing lists, contact the DMA. [See the "Additional Help and Information" section of this *Sourcebook* for contact information.]

4. Ask the credit bureaus not to share your personal information: The major credit bureaus offer an "opt-out" choice that limits the information shared with others or used for promotional purposes. When you "opt-out," you may cut down on the number of unsolicited telemarketing calls that you receive. Remember that if you "opt-out," you also will reduce the number of offers of credit that you receive by mail. The credit bureaus offer a toll-free number to call to "opt-out" of having pre-approved credit offers sent to you for two years. Call 1-888-5-OPTOUT (567-8688) for more information.

In addition, you can notify the three major credit bureaus that you do not want personal information about you shared for promotional purposes. [See the "Additional Help and Information" section of this *Sourcebook* for contact information.]

You Make the Call: The FTC's New Telemarketing Sales Rule

[Editor's Note: This information update was released on December 18, 2002 by the FTC.]

The Federal Trade Commission's amended Telemarketing Sales Rule (TSR) will put consumers in charge of the number of telemarketing calls they get at home. With the creation of a national "do not call" registry, the FTC will make it easier and more efficient for consumers to stop getting the telemarketing sales calls they don't want.

One caveat: Once funding is available, it will take about four months to create the registry, another two months for consumers to enroll region by region, and then another month before telemarketers are required to search the registry and honor the wishes of consumers who don't want to be called. Until the national "do not" call registry is up and running, consumers can limit the calls they get by telling callers to put them on their company's own "do not call" list.

When Will the National "Do Not Call" Registry Take Effect?

The FTC will begin creating the national "do not call" registry as soon as funding is available. Consumers will be able to put their names on the registry online or by telephone about four months later. An estimated three months after that, telemarketers will be required to search the registry. That's when consumers who have placed their home telephone numbers on the national "do not call" registry should notice a difference in the number of telemarketing calls they receive. And that's when enforcement efforts against those firms who do not honor the registry will begin.

The FTC expects that many consumers will want to put their telephone numbers on the national "do not call" registry. To facilitate the process, registration will be phased in, region by region. Check www.ftc.gov/donotcall for details on when consumers in your state can sign up.

How Will the National "Do Not Call" Registry Work?

You will be able to register for free online or by calling a toll-free number. If you are registering by phone, you will have to call from the telephone number that you wish to register. If you register online, you may need to provide limited personal information for confirmation. The only identifying information that will be kept in the registry will be the phone number you register. You can expect fewer calls within three months of the date you sign up for the registry.

Your number will stay in the registry for five years or until you take your number out of the registry or change phone numbers. After five years, you will be able to renew your registration.

The law requires telemarketers to search the registry every 90 days and delete from their call lists phone numbers that are on the registry. If you find that you are receiving telemarketing calls even after you have registered your telephone number, you will be able to file a complaint with the FTC online or by calling a toll-free number. A telemarketer who disregards the national "do not call" registry could be fined up to $11,000 for each call.

Who Is Covered by the National "Do Not Call" Registry?

Placing your number on the national "do not call" registry will stop most, but not all, telemarketing calls. Some businesses are exempt from the TSR and can still call you even if you place your number on the registry. (These include long-distance phone companies and airlines, and insurance companies that operate under state regulations.)

But most telemarketing calls are placed by professional telemarketing companies, and they are not exempt, even if they are calling on behalf of an exempt company. The bottom line: Professional telemarketers cannot call you if you are on the registry.

There are other business that are not required to "go by the list." For example, organizations with which you have an established business relationship can call you for up to 18 months after your last purchase, payment or delivery—even if your name is on the national "do not call" registry. And companies to which you've made an inquiry or submitted an application can call you for three months. However, if you ask a company not to call you, it must honor your request, even if you have an established business relationship.

If you place your number on the national registry, you may give written permission to particular companies that you want to hear from. If you don't put your number on the national registry, you can still prohibit individual telemarketers from calling, one by one, by asking them to put you on their company's "do not call" list.

One more important point: Although callers soliciting charitable contributions do not have to search the national registry, a for-profit telemarketer calling on behalf of a charitable organization must honor your request to be put on its "do not call" list.

How Does the National "Do Not Call" Registry Square with State Lists?

Many states have "do not call" registries. The FTC is working to coordinate the national "do not call" registry with these states to avoid duplication. This process will take a year or more; check the FTC's website or your state attorney general's office for details.

Are There Other Protections against Unwanted Telemarketing Calls?

The Telemarketing Sales Rule prohibits deceptive and abusive telemarketing acts and practices and protects you from unwanted late-night telemarketing calls:

- Calling times are restricted to the hours between 8 a.m. and 9 p.m.
- Telemarketers must promptly tell you the identity of the seller or charitable organization and—before they make their pitch—that the call is a sales call or a charitable solicitation.

- Telemarketers must disclose all material information about the goods or services they are offering and the terms of the sale. They are prohibited from lying about any terms of their offer.

In addition to creating the national "do not call" registry, the amendments to the TSR will:

- **Greatly reduce abandoned calls.** Telemarketers will be required to connect the call to a sales representative within two seconds of the consumer's greeting. This will reduce the number of "dead air" or hang-up calls you receive from telemarketers. These calls result from the use of automatic dialing equipment that sometimes reaches more numbers than there are available sales representatives. In addition, when the telemarketer doesn't have a representative standing by, a recorded message must play to let you know who's calling and the telephone number they're calling from. The law prohibits a sales pitch. And to give you time to answer the phone, the telemarketer may not hang up before 15 seconds or four rings.
- **Restrict unauthorized billing.** Before billing charges to your credit card account, telemarketers will be required to get your express informed consent to be charged—and to charge to a specific account. If a telemarketer has your account information before the call and offers you goods or services on a free trial basis before charging you automatically—also known as a "free-to-pay conversion" offer—the telemarketer must get your permission to use a particular account number, ask you to confirm the number by repeating the last four digits, and, for your protection, record the entire phone transaction.
- **Require Caller ID transmission.** Telemarketers will be required to transmit their telephone number and if possible, their name, to your Caller ID service. That will protect your privacy, increase accountability on the part of telemarketers, and help in law enforcement efforts. This provision will take effect one year after the release of the Rule.

Chapter 29

"No-Call Law" Hangups

They've got your number and they're going to use it, by golly. Despite Zappers, Caller ID boxes, telemarketer voodoo dolls, and even "Do Not Call" laws, they will call every night at 6:34 sharp during the macaroni and cheese course to pitch their wares.

Why? Because they can. Legally. The "No-Call List Act," which is in place in more than 20 states, has so many exemptions that most consumers who put their name on the list get little to no relief from phone solicitations. For example, political campaigns are exempt, as are calls from charitable, religious, and nonprofit organizations. If a salesman wants to arrange a face-to-face meeting, he's got your number and he's legally allowed to use it. Same goes for the financial services providers with whom you do business. Your bank can ring you up at any time to pitch you their new credit card/Rotato combo offer.

According to a study by the American Teleservices Association in Washington (to protect worker safety we will not divulge the address here), 65% of telemarketing calls are made on behalf of political groups or charities—two categories exempted from the lists.

It's a Catch-22. "Do Not Call" laws are very popular with voters. But not so much with politicians who rely heavily on telemarketing to raise campaign funds and get voters to the polls. And there's the rub: Do away with the exemptions and the politicos cut off a big source

---

""No-Call Law" Hangups," *The Motley Fool*, http://www.fool.com/news/take/2002/take020314.htm#no, March 14, 2002.

of campaign funding. Speak out against the No-Call List Act and they lose favor with voters.

Now the Federal Trade Commission (FTC) is stepping in with proposed changes to the Telemarketing Sales Rule, including the creation of a centralized national "Do Not Call" registry. Though it still doesn't block all solicitors from calling, it appears to carry fewer exemptions.

This article first appeared in The Motley Fool's March 14, 2002 *Fool Watch* newsletter.

Chapter 30

Predictive Dialing: Silence on the Other End of the Line

The Federal Communications Commission (FCC) receives many complaints about predictive dialing by telemarketers. Here's what happens: the phone rings and when the person receiving the call picks up the phone, he or she is met with silence or the "click" of the calling party disconnecting the call. This can be caused by predictive dialing, which is when a telemarketer's automatic dialer simultaneously dials many more numbers than the telemarketer can handle if all of the called parties pick up at the same time. The first to pick up is connected to the telemarketer while the rest are disconnected.

FCC rules require that companies identify themselves to consumers and also that telemarketers maintain "do-not-call" lists for people who do not wish to receive telemarketing calls from a certain company. The practice of predictive dialing, and the resulting abandoned calls, often do not allow consumers to identify the company calling and, therefore, do not afford consumers the opportunity to make a "do-not-call" request under FCC rules.

Consumers who wish to avoid all telemarketing calls may want to contact their state to find out if it has a broad "do-not-call" law that prohibits all telemarketing calls to individuals registered on its state list. Currently, Alabama, Alaska, Arkansas, California, Colorado, Connecticut, Florida, Georgia, Idaho, Illinois, Indiana, Kentucky, Louisiana, Maine, Missouri, Oregon, New York, Tennessee, Texas and Wyoming

---

Federal Communications Commission (FCC), http://www.fcc.gov/cgb/consumerfacts/PredictiveAlert.html, reviewed/updated March 25, 2002.

have implemented, or are in the process of implementing, "do-not-call" lists. Other states are considering similar laws. Also, the Direct Marketing Association (DMA) maintains a "do-not-call" list that is used voluntarily by its 4,500 member companies.

For more information on predictive dialing, contact your state utility commission or consumer protection agency.

Chapter 31

Telemarketing Recovery Scams

Scam artists buy and sell "sucker lists" with the names of people who have lost money in fraudulent promotions. These so called "recovery room" operators may promise to recover the money you lost or the prize or merchandise you never received—for a fee that they want you to pay in advance. That's against the law. Under the Telemarketing Sales Rule, it is illegal for recovery room operators to request or receive payment until seven business days after they deliver the recovered money or other item to you. If a recovery operation asks you to pay up front, don't do business with them!

How the Scams Work

Consumers who have lost money through prize promotions, merchandise sales, and charity drives often are placed on "sucker lists." These lists contain names, addresses, phone numbers, and other information, such as how much money the consumer has spent responding to telemarketing solicitations. "Sucker lists" are bought and sold by unscrupulous promoters who know that consumers who have been deceived once are vulnerable to additional scams.

When recovery room scam operators contact consumers who have lost money to a previous telemarketing scam, they falsely promise

---

Federal Trade Commission (FTC), http://www.ftc.gov/bcp/conline/pubs/tmarkg/recovery.htm, May, 1996. Despite the age of this document, readers seeking information on telemarketing recovery scams will find this information useful.

that, for a fee or donation to a specified charity, they will recover the lost money, or the prize or product never received. The operators use a variety of misrepresentations to add credibility to their pitch. Some claim to represent companies or government agencies. Some say they are holding money for you. Others offer to file necessary complaint paperwork with government agencies on your behalf. Still others claim they can get you placed at the top of a list for victim reimbursement.

The Federal Trade Commission (FTC) repeatedly finds claims like these to be false. Although some federal and local government agencies and consumer organizations provide assistance to consumers who have lost money, they do not charge a fee, guarantee to get back even a portion of your money, or give special preference to people who file formal complaints. The fact is that many recovery room scam operators never file your complaint. But they may charge exorbitant fees for providing addresses of pertinent government agencies—addresses that nonprofit consumer protection agencies may offer free-of-charge or at a low cost. It is very difficult to get your money back; the fraudulent telemarketer who contacted you initially is not likely to have the money readily available or even still be in business.

Protecting Yourself

To avoid losing money to a "recovery room" operator, take the following precautions:

- Be skeptical of people who offer to recover money, merchandise, or prizes, especially those who want their fee in advance. Under the FTC's Telemarketing Sales Rule, it is illegal for a recovery room operator to request or receive payment until seven business days after you receive the recovered money or other item.

- Beware of individuals claiming to represent companies, consumer organizations, or government agencies that will recover your lost money, merchandise, or prizes for a fee or donation to a special charity. National, state, and local consumer enforcement agencies and nonprofit organizations do not charge for their services. These groups include your state Attorney General (AG) and local consumer protection office.

- Before you purchase any recovery services by phone, ask what specific services the company provides and the cost of each service.

Ask the company to send you written materials about its operation. And, check out the company with local law enforcement and consumer agencies. Ask if other consumers have registered complaints about the business.

- Don't give out your credit card or checking account numbers unless you have a relationship with the business or know its reputation.

Chapter 32

Fraudulent Telemarketers Snatching Bank Account Numbers

In 2001 consumers paid for fraudulent telemarketing offers more frequently with bank debits than any other payment method, according to statistics released today by the National Consumers League's National Fraud Information Center (NFIC). Bank debits are situations in which fraudulent telemarketers obtain consumers' bank account numbers either by luring consumers into providing them or getting them from another source. Bank debits are particularly high in certain categories among the top ten telemarketing frauds in 2001: 62 percent of consumers paid for bogus credit card offers with bank debits; 50 percent gave out their bank numbers in the increasingly popular age-old Nigerian money offer scam; and 26 percent had funds withdrawn from their bank accounts for buyers' clubs they never agreed to purchase.

"People should guard their personal financial information carefully, because it may be hard to prove that they never authorized the withdrawals and get that money back in their accounts," says Susan Grant, director of NFIC. She notes that the Federal Trade Commission is proposing to amend the federal Telemarketing Sales Rule to outlaw companies from getting consumers' billing information from or providing it to third parties.

"Fraudulent Telemarketers Snatching Bank Account Numbers, National Fraud Information Center Stats Show," National Consumers League, http://www.natlconsumersleague.org/telepr0202.htm, February 6, 2002. Used with permission.

National Consumers League (NCL) launched a new consumer education initiative in February 2002, "Stop Calling Me! How to Get Off Marketing Lists," to help consumers take charge of their privacy. A new section of NCL's Web site, www.nclnet.org/privacy, has been unveiled in connection with National Consumer Protection Week 2002, which focuses on privacy.

"Stop Calling Me!" features tips about how to avoid getting on marketing lists, how to remove your name, and what to do if your privacy is violated. Links are provided to make it easy for consumers to get more information and take action. There is a brochure and a tip sheet that can be downloaded and printed to use as handouts. This new section of NCL's Web site also offers other useful information for consumers, such as "Online E-ssentials," a consumer guide to Internet privacy and security.

"Invasions of privacy can be more than just annoying," says Grant. "Your personal information can be abused for illegal purposes." She notes that fraudulent telemarketers use lists to target likely victims. For example, seniors are disproportionately targeted for sweepstakes scams, with people 60 and older representing 60 percent of victims in that category in 2001, despite making up only 26 percent of telemarketing victims overall.

2001 Top Telemarketing Frauds

The top ten telemarketing frauds reported to NFIC are, in order: work-at-home schemes, prizes/sweepstakes, credit card offers, advance fee loans, magazine sales, telephone slamming, buyers' clubs, credit card loss protection, Nigerian money offers, and telephone pay-per-call services.

Tips to Get off of Marketing Lists

- Don't provide information that isn't necessary for the transaction.
- Be anonymous. Consider using online tools and fictitious names in situations where your real identity isn't needed and there is no other option to avoid getting on marketing lists.
- Think twice before entering contests.
- Know the privacy policy.
- Understand your privacy choices. If there is no privacy policy or it doesn't allow you to avoid unwanted marketing, take your business elsewhere.

- Know when your personal information is being collected. Be aware of Automatic Number Identification and other ways that your information may be collected, and tell the company if you don't want to be put on a marketing list.
- Understand that unlisted and unpublished phone numbers don't guarantee privacy.
- Know your telemarketing rights. Federal law allows you to tell marketers not to call you again. Check with your state attorney general's office to find out if you also have "Do Not Call" rights under state law.
- Know your financial privacy rights. Federal law requires financial institutions to tell you what information they collect and how they use it and allows you to request that your personal information not be shared with unrelated companies. Check with your state attorney general's office to find out if you also have financial privacy rights under state law.
- Know your medical privacy rights. Federal regulations limit how your health information can be used and shared with others for marketing purposes. Check with your state attorney general's office to find out if you also have medical privacy rights under state law.

Your state may protect you against "spam." Some states have enacted laws about unsolicited e-mails. Check with your state attorney general's office.

The National Consumers League, founded in 1899, is America's pioneer consumer organization. Our mission is to identify, protect, represent, and advance the economic and social interests of consumers and workers. NCL is a private, nonprofit membership organization. For more information, visit www.nclnet.org.

Chapter 33

Inbound Telemarketing Fraud

In what may be the most far-reaching Federal Trade Commission (FTC) law enforcement sweep ever against "in-bound" telemarketing fraud—where consumers call companies based on classified ads, Internet banners, or other promotions—the FTC announced the filing of 11 federal district court complaints in April 2002. Among those charged were the purveyors of advance-fee loans and credit cards, at-home medical billing programs, work-at-home envelope stuffing schemes, and a "consumer protection" agency that was, in reality, no more than a shill for a vending machine business opportunity.

In each case brought through "Operation Dialing for Deception" the Commission charged the defendants with violating the FTC Act, the Telemarketing Sales Rule (TSR), or both. In all 11 complaints, the FTC is either seeking—or has received—relief ranging from temporary restraining orders to preliminary or permanent injunctions, as well as a freeze of the defendants' assets and the appointment of a receiver to oversee their finances pending trial, where appropriate.

"Consumers spotting a classified ad or telephone pole promotion and acting on their curiosity by calling the number face the same risk of being misled, deceived, or defrauded as they do when responding to a high-pressure sales call from a telemarketer," said FTC Bureau of Consumer Protection Director Howard Beales III. "If the offer appears too good to be true, be on guard."

Excerpted from "FTC Sweep Protects Consumers from 'Dialing for Deception,'" Federal Trade Commission (FTC), http://www.ftc.gov/opa/2002/04/dialing.htm. April 15, 2002.

"White collar crime is on the increase and a favorite weapon of choice is the telephone. Don't abandon common sense just because you initiated the call, and it would be wise to check with the Better Business Bureau first," Ken Hunter, president and CEO of the Council of Better Business Bureaus, advised consumers.

A Focus on Medical Billing Scams

Beales said that in addition to the general message that consumers should be careful when calling companies in response to classified ads or similar promotions, five of the cases filed by the FTC illustrate the fact that consumers interested in working at home doing medical billing should be particularly wary of pitches that promise easy money with little or no effort. While the telemarketers may provide lists of local doctors they say are interested in having their billing done by consumers, many times these doctors have not consented to have their information distributed, are not looking for outside help, or may need more skilled employees to complete this technical task.

"I would advise someone looking to start his or her own medical billing business to learn about the challenges involved in medical billing, including the complex laws which apply to [it]," said Cyndee Weston, Executive Director of the American Medical Billing Association. "I would also advise them that up to one year of training may be necessary in order to even begin to market his or her medical billing services to healthcare providers."

In 2001, the FTC and 43 Better Business Bureaus (BBBs) across the United States and Canada surfed the Internet and newspaper classifieds looking for ads promising consumers they could make fast, easy money running medical billing businesses from home. Hundreds of ads from dozens of companies were identified, with the worst purveyors of medical billing fraud targeted by the five cases announced in April 2002. The remainder received warning letters that their practices may be in violation of federal law.

Readers may obtain further information on the individual cases by visiting the Federal Trade Commission's website at www.ftc.gov/opa/2002/04/dialing.htm.

Chapter 34

Telephone Harassment: How to Cope with the Crank Caller

The phone rings. It's 3 a.m. No, it's not Mom calling to say Dad just deposited $500 in your overdrawn checking account. Instead you are greeted by a string of obscenities, heavy breathing or silence.

You have just become another victim of the crank caller. Some authorities estimate that over a million crank calls were made last year. Most of these calls are the ring-and-hang-up variety. A growing number are either obscene or threatening.

Now that you've become the victim of the crank caller, here are a few ideas to prevent a repeat call.

1. Hang up. As soon as you hear an obscenity, improper questions or no response to your sleepy "hello?"—HANG UP!!

2. Don't talk to strangers. Be careful when the caller says he/she is taking a survey. If you have any concern over the legitimacy of the survey, ask the person for his/her name, firm name and telephone number. Say that you will call back after you verify the authenticity of the survey.

3. Don't play detective. Don't extend the call trying to figure out who is calling. This or any type of reaction is exactly what the caller wants and needs.

"Telephone Harassment: How to Cope with the Crank Phone Caller," Ohio University Police Department, http://www.ohiou.edu/police/safephone.htm. Downloaded April 16, 2002. Ohio University Police Department, © 2002.

4. Keep cool. Don't let the caller know you are upset or angry.

5. Don't panic when the phone rings. It may just be a wrong number. If it is a crank call, follow the suggestions here.

6. Don't try to be clever. A witty response may well be interpreted as a sign of encouragement.

7. Don't try to be a counselor. The annoyance or obscene caller certainly needs professional help, but he or she will only be encouraged by your concern and will continue the late night calls.

8. Don't tell everyone about your calls. Many calls of this type are actually made by friends, family members, even your closest friend or someone you are dating.

9. Place ads with caution. When placing an ad in a newspaper, use a newspaper or post office box number if possible. If you must use your phone number, do not list your address. Crank callers are avid readers of the classified ads.

10. Never volunteer your number to an unknown caller. This is an invitation to call again. If your number is the wrong number, the caller does not need to know your number.

11. Report obscene or annoying calls to the Ohio University Police Department if they are coming to a campus phone. A crank call log can be obtained from the police department. Use it to make a record of the call. By reporting the call to the above authorities, you have begun the process to stop these 3 a.m. calls.

Chapter 35

Telephone Harassment: Using Caller ID, Trap and Trace, and Other Methods

Obscene or harassing phone calls can be one of the most stressful and frightening invasions of privacy a person experiences. And unwanted phone calls, while a minor problem when compared with threatening calls, can still be a major inconvenience. Fortunately, there are steps you can take to help put an end to these unwelcome intrusions.

What Makes a Phone Call Harassing?

When someone calls and uses obscene or threatening language, or even heavy breathing or silence to intimidate you, you are receiving a harassing call. It is against the law in California and other states to make obscene or threatening calls. (California Penal Code section 653m)

How Often Do I Have to Get These Calls to Make It Harassment?

Just one unwelcome call can be harassing; but usually your local phone company will not take action unless the calls are frequent.

"Fact Sheet 3: Stopping Harassing Phone Calls: How to Put an End to Unwanted or Harassing Phone Calls," Privacy Rights Clearinghouse, http://www.privacyrights.org/fs/fs3-hrs2.htm. Reprinted with permission from the Privacy Rights Clearinghouse, a non-profit consumer advocacy and information program located in San Diego, CA. © 2002. Contact the Privacy Rights Clearinghouse at 3100-5th Ave., Suite B, San Diego, CA 92103, (619) 298-3396 (voice), (619) 298-5681 (fax), prc@privacyrights.org (E-mail). For the most recent version of this fact sheet, or additional information, visit www.privacyrights.org.

However, if a call specifically threatens you or your family with bodily harm, the phone company will generally take immediate action.

Who Should I Contact When I Get Harassing Calls?

Local phone companies have varying policies on whether to call the phone company or the police first. Some recommend that you first call the phone company's business office and explain the problem. A representative will connect you with the "annoyance desk." Other phone companies may require you to file a formal complaint with local law enforcement before they will deal with the matter. To find out what your phone company's policy is, contact the business office and ask for assistance.

For serious threats, if life or property are threatened, or if calls are obscene, you should call the police and file a report. Provide as much information to law enforcement as you can. Indicate the gender of the caller and describe the caller's voice. Note the time and date of the call(s). What did the caller say? How old did he/she sound? Did the caller seem intoxicated? Did he/she have an accent or speech impediment? Was there any background noise? Was a phone number/name displayed on the Caller ID device?

What Can My Local Phone Company Do If I Am Receiving Harassing Calls?

If the calls are frequent or particularly threatening, the phone company can set up a **"Trap"** on your phone line. The Trap allows the phone company to determine the telephone number from which the harassing calls originate. You must keep a log noting the time and date the harassing calls are received. Traps are usually set up for no more than two weeks. The phone company does not charge a fee for Traps.

A phone company service called **Call Trace** may also be able to help track down harassing calls. Immediately after receiving a harassing call, you enter the code *57 on your phone and the call is automatically traced (1157 on rotary phones). Call Trace is easier than using a Trap since the customer does not have to keep a phone log. But Call Trace technology works only within the local service area. (Look in the "Customer Guide" section of the phone book or the phone company's web site for a description of your local service area.)

Call Trace must be set up in advance by the individual receiving harassing calls, and it requires a fee for use. However, in situations where the phone company would ordinarily use a Trap, you might not be charged if the phone company suggests that Call Trace be used as an alternative. Be sure to ask.

The information collected from Call Trace or from a Trap is turned over to law enforcement personnel, not the customer. Law enforcement officers try to stop the harassing calls by either warning or arresting the harasser. With both Call Trace and a Trap, your phone conversations are not listened to or recorded by the phone company.

Is the Phone Company Always Able to Solve Harassing Phone Call Problems?

No. If the caller uses a phone booth or multiple phone lines, the phone company and law enforcement officials may never get enough identification to take further action. In cases like these, changing your phone number might help. Also, you might want to get an unlisted or unpublished number. In addition, the tips listed below for discouraging other types of unwanted calls may be of help.

What Can I Do to Stop Harassing Calls Without Going to the Phone Company or Police?

First, simply hang up on the caller. Do not engage in conversation. Typical crank callers are seeking attention. You have "made their day" if you say something to them or express shock or anger.

If the silent treatment does not work, you might try putting a message like this on your voice mail system:

> *I'm sorry I/we can't come to the phone right now but you must leave a message. I/we are receiving annoyance calls and the phone company has a trap on this line. If you do not leave a message I/we will assume that you are the annoyance caller and this call will be traced.*

If you answer the phone and the harassing caller is on the line, another suggestion is to say: "Operator, this is the call." Then hang up. Or say the word "trap," what time it is and the date; then hang up.

What Is the "Pressure Valve" Strategy?

Some threatening calls are part of a larger pattern of abuse, such as stalking. Some experts recommend in these situations to get a new phone number, but keep the phone number being called by the harasser and attach a voice mail machine or message service to that line. Turn the phone's ringer off and don't use that phone line for anything other than capturing the calls of the harasser.

This is the pressure valve strategy. The harasser will continue to call the unused number and will think that he/she is getting through. Instead, you are simply using the number to gather evidence. You will want to save tape recordings of the calls.

Get another phone number for your use, and be sure it's unlisted and unpublished. Give the number to trusted friends and relatives only. Do not give it to your bank, credit card company or credit bureau. Put passwords on all of your phone accounts (local, long distance, and mobile). Tell the phone companies in writing that they must not disclose any account information to anyone but yourself, and only when the correct password is given.

What Precautions Can I Take to Prevent Harassment?

Do not disclose personal information when called by someone you do not know. They might be checking out the residence for possible robbery or other crime. If the caller asks what number they have called, do not give it. Instead, ask them to tell you what number they dialed.

To prevent being targeted for obscene calls and heavy breathing, women should only list their first initial and last name in the phone directory. Having an unlisted number is another option.

Children should be instructed to never reveal information to unknown callers. Instead, they should be taught to record the caller's name and phone number along with date and time.

Do not include your telephone number on the outgoing message of your voice mail service if you wish to keep your number private. By omitting your phone number from your message, you prevent random dialers and people with Call Return (explained below) from capturing this information.

Sometimes My Phone Rings and There Is No One on the Line. What Is Happening?

Many people are frightened when they receive "hang-up" calls. They wonder if someone is harassing them, or if a burglar is checking to see if they are not home. In most cases, these calls are from telemarketers. (For additional information on telemarketing, see Fact Sheet No. 5, www.privacyrights.org/fs/fs5-tmkt.htm.)

Many telemarketers use "predictive dialing" technology to call consumers. A computer dials many phone numbers in a short period of time. When an individual answers, the computer seeks a sales representative who is not occupied at that time and connects the call. If all of the sales reps are on calls, the consumer hears dead silence. These are "abandoned calls."

Several devices claim to stop these calls, including Telezapper (www.telezapper.com) available in stores that sell consumer electronics, and Call-Me-Not (www.callmenot.com) at (800) 860-8465. (No endorsements are implied.)

If you are receiving many abandoned calls a day, you can call the annoyance department of your local phone company and ask that a Trap be placed on your line. In extreme situations, the phone company might be willing to contact the offending telemarketer and request that your phone number be place on its "do not call" list. If the repeated calls are from a malicious individual who is harassing you rather than a telemarketer, the phone company will report the number to law enforcement as described in the beginning of this guide.

A new California law requires telemarketers to limit abandoned calls to fewer than 1% of their total call volume effective January 2003. For information on California Public Utilities Code 2875.5, visit www.leginfo.ca.gov/calaw.html.

What Can I Do to Stop Other Kinds of Unwanted Calls?

Sometimes calls are annoying but are not serious enough to involve law enforcement as is necessary with either a Trap or Call Trace. These might include telemarketing sales calls, wrong numbers, overly aggressive bill collectors, and prank calls. There are several steps you can take to discourage such unwanted calls.

An **answering machine** or a **voice mail service** is one of the best ways to limit unwanted calls. Available for as little as $30, an answering machine records messages when you are not available and can also be used to screen your calls. Similar to an answering machine, a voice mail service or an answering service can also discourage unwanted calls.

Another product on the market is an attachment to the telephone called an **"inbound call blocker."** It allows only those callers who enter a special numeric code onto their touch tone phone pad to ring through to your number. This device is highly effective in preventing unwanted calls. However, you must be certain to give the code to everyone you want to talk to. Even so, you could miss important calls from unexpected sources, like emergency services.

Several vendors sell such call screening devices. Check the web site of Privacy Corps (www.privacycorps.com) or call (888) 633-5777. Other sources include Command Communications (www.command-comm.com), at (800) 288-3491; and Avinta (www.avinta.com) at (800) 227-1782. No endorsements are implied.

In most areas of the country, **Custom Calling** services are available from the local phone company which can help limit unwelcome calls. However, before you sign up, look carefully at the services to be certain they will work in your situation and are worth the monthly fee.

Also remember that many of these features only work within your local service area. Calls coming from outside the area might not be affected by these features. (Consult the "Customer Guide" section of the phone book or the company's web site to find out the boundaries of your local service area.) Keep in mind, these services require a fee, either month-to-month or per-use. To avoid having to pay for call screening on an ongoing basis, consider purchasing a device that attaches to the telephone, such as the call screening devices mentioned above.

*Call Screen (*60)*

Your phone can be programmed to reject calls from selected numbers with a service called Call Screen (SBC Pacific Bell term; other phone companies might use a different name). Instead of ringing on your line, these calls are routed to a recording that tells the caller you will not take the call. With Call Screen, you can also program your telephone to reject calls from the number of the last person who called. This allows you to block calls even if you do not know the phone number. Most phone companies charge a monthly fee for this service.

Call Screen is not a foolproof way to stop unwelcome calls. A determined caller can move to a different phone number to bypass the block. Also, Call Screen does not work on long distance calls from outside your service area.

Priority Ringing

You can assign a special ring to calls from up to 10 numbers the calls you are most likely to want to answer. The rest can be routed to voice mail. There are ways callers can get around Priority Ringing when it is used as a screening tool. The harasser can switch phone lines and avoid the distinctive ring.

*Call Return (*69)*

This service allows you to call back the number of the last person who called, even if you are unable to answer the phone. Some people suggest that Call Return can be used to stop harassing callers by allowing you to call the harasser back without knowing the phone number. Use caution with this method of discouraging harassing callers, however, as it could actually aggravate the problem. This service is paid on a per-use basis.

Can I Use Caller ID to Stop Unwanted Calls?

With **Caller ID**, customers who pay a monthly fee and purchase a display device can see the number and name of the person calling before picking up the phone. Some people believe Caller ID will help reduce harassing or unwelcome calls. Others, however, raise privacy concerns about the technology since subscribers to the service can capture callers' phone numbers without their consent.

To help consumers protect the privacy of their phone numbers, state public utilities regulators (for example, the California Public Utilities Commission) require local phone companies to offer number blocking options to their customers.

There are two blocking options to choose from. If the customer chooses **Per Line Blocking** (called **Complete Blocking** in California), their phone number will automatically be blocked for each call made from that number. If the customer chooses **Per Call Blocking** (called **Selective Blocking** in California), the phone number is sent to the party being called unless *67 is entered before the number is dialed. When the number is blocked by either of these methods, the Caller ID subscriber sees the word "private" or "anonymous" on the Caller ID display device.

Because of these blocking options, Caller ID is not likely to allow you to capture the phone number of the determined harasser. Most harassers will block their phone numbers or will call from pay phones. However, Caller ID can be used by people receiving harassing calls to decide whether to answer. They can choose not to pick up calls marked "private" or numbers they don't recognize.

A companion service to Caller ID, called **Anonymous Call Rejection (ACR)**, requires an incoming call from a blocked number to be unblocked before the call will ring through. Use of this feature forces the harasser to disclose the number by entering *82 or to choose to not complete the call. But a determined harasser can get around this feature by using a pay phone. This service can be added to a consumer's local phone service for a fee or at no charge depending on the carrier. It is activated and deactivated with the touch tone code *77.

What Does Privacy Manager Do?

Most local phone companies offer a relatively new service called **Privacy Manager**. It works with Caller ID to identify incoming calls that have no telephone numbers. Calls identified as "anonymous," unavailable," out of area" or "private" must identify themselves in order to complete the call. Before your phone rings, a recorded message instructs the caller to unblock the call, enter a code number (like the inbound call blocking devices mentioned above), or record their name. When your phone rings, you can choose to accept or reject the call, send it to voice mail, or send a special message to telemarketers instructing them to put you on their "do not call" list.

Chapter 36

Unwanted Faxes: What You Can Do

Background

A telephone facsimile, or "fax" machine is able to send and receive data (text or images) over a telephone line. The Telephone Consumer Protection Act of 1991 (TCPA) and Federal Communications Commission (FCC) rules prohibit sending unsolicited advertisements, also known as "junk faxes," to a fax machine. This prohibition applies to fax machines at both businesses and residences.

Definitions

An "unsolicited advertisement" is defined as "any material advertising the commercial availability or quality of any property, goods, or services which is transmitted to any person without that person's prior express invitation or permission." Just because your fax number is published or distributed does not mean others have permission to send you unsolicited advertisements. If, however, you have an "established business relationship" with a person or entity then, in effect, you've given your consent to receive unsolicited faxes from that person or entity.

You have an "established business relationship" with a person or entity if you have made an inquiry, application, purchase, or transaction regarding the products or services offered by that person or entity.

Federal Communications Commission (FCC), http://www.fcc.gov/cgb/consumerfacts/unwantedfaxes.html, reviewed/updated March 25, 2002.

If that person or entity has been sending you fax advertisements and you want them to stop, you can end the established business relationship by telling whoever is sending the faxes that you do not want any more unsolicited advertisements sent to your fax machine. The transmission of unsolicited advertisements by a person or entity with whom you no longer have a business relationship would violate the TCPA.

How the FCC Can Help

The FCC has taken numerous enforcement actions, including citations and fines, against companies for violations and suspected violations of the TCPA's prohibition against unsolicited faxes. Consumers who have received unsolicited faxes are encouraged to contact the FCC regarding the incident(s). You may need to provide documentation in support of your complaint, such as copies of the fax(es) you received.

Additional Places to Go for Help

You can also file TCPA-related complaints with your state authorities, including your local or state consumer protection office or your state Attorney General's office.

It is also possible to bring a private suit against the violator in an appropriate court of your state. Through a private suit, you can either recover the actual monetary loss that resulted from the TCPA violation, or receive up to $500 in damages for each violation, whichever is greater. The court may triple the damages for each violation if it finds that the defendant willingly or knowingly committed the violation.

Part Five

Wiretapping and Electronic Eavesdropping

Chapter 37

Wiretapping/Eavesdropping on Telephone Conversations: Is There Cause for Concern?

Overview of the Threat

A bug is a device placed in an office, home, hotel room, or other area to monitor conversations (or other communications) and transmit them out of that area to a listening post. Other listening devices work from a distance to monitor communications within a room without actually having a microphone or transmitter in the room.

Thanks to an explosion of miniaturized technology, the tools for bugging and other forms of eavesdropping have never been cheaper, smaller, more powerful, or easier to come by. By one account, $888 million worth of eavesdropping devices are sold in the United States each year.[1] A leading technical security countermeasures expert has said, "I can't drive more than four blocks in any direction in midtown Manhattan without picking up an eavesdropping device."[2]

Spy paraphernalia now ranges from supersensitive microphones hidden in pens to video cameras that will fit behind a tie and take pictures through the tieclip. Bugging devices can be made to look like,

---

Excerpted from "Bugs and Other Eavesdropping Devices," from an undated website produced by the Counterintelligence Training Academy, Nonproliferation and National Security Institute, Department of Energy (DOE), http://www.smdc.army.mil/Intelligence/Security%20Guide/V3bugs/overthreat.htm. Downloaded March 2002; and "Eavesdropping Methods," from an undated website produced by the Counterintelligence Training Academy, Nonproliferation and National Security Institute, Department of Energy (DOE), http://www.smdc.army.mil/Intelligence/Security%20Guide/V3bugs/Methods.htm. Downloaded March, 2002.

and actually function as, fountain pens, clock radios, desk calculators, telephone jacks, and even teddy bears.[3]

The widespread availability of tools for covert surveillance represents a threat to national security information, law enforcement and other government operations, the confidentiality of business transactions, and to personal privacy.

The different types of eavesdropping paraphernalia and how they are used are discussed below under "Eavesdropping Methods." This does not discuss the various devices in detail—just enough to provide a realistic understanding of your vulnerabilities. The goal is to encourage you and help you to protect information by:

- Being more careful to protect the physical security of your office and home to prevent others from gaining access to install eavesdropping devices.
- Being more careful where you talk. Conversations can be bugged even in public areas such as airplanes, restaurants, hotel lobbies, and public parks.
- Being more alert to evidence that your conversations may have been overheard. One company gradually lost its competitive advantage to the point that it went bankrupt. When the furniture was being moved out of the office, an active microphone was found behind a large credenza in the conference room. The company suspects the microphone may be at least partially responsible for its bankruptcy.

Recognizing this threat, Congress passed the Electronic Communication Privacy Act of 1986. This law makes it illegal for private citizens to own, manufacture, import, sell or advertise any eavesdropping device while "knowing or having reason to know that the design of such device renders it primarily useful for the purpose of the surreptitious interception of wire, oral, or electronic communications, and that such device or any component therefor has been or will be sent through the mail or transported in interstate or foreign commerce." The penalty for violation is a fine of up to $10,000 and/or prison for up to five years. According to the act, only law enforcement authorities are allowed to use these "bugs," and then only after obtaining a court order. In 1993, Federal agents raided "spy shops" in 24 cities to shut down the widespread illegal marketing of surveillance devices.

Despite the legal sanctions, sales of surveillance devices that may violate the law are open and widespread in the United States. Technical surveillance is a common espionage tool and a common source of competitive information in industry.

It is noteworthy that legal restrictions against technical surveillance have not been adopted by other countries with which we have close trading ties, so American companies engaged in international commerce are particularly vulnerable. Although industrial espionage is unlawful, the rewards for procuring intelligence regarding the strategic plans, resources, products, pricing, customers, personnel, or legal affairs of a competitor often prove substantially more persuasive than concern over the risk of being caught acquiring such information.

Eavesdropping Methods

Eavesdropping operations generally have three principal elements:

- **Pickup Device:** A microphone, video camera or other device picks up sound or video images and converts them to electrical impulses. If the device can be installed so that it uses electrical power already available in the target room, this eliminates the need for periodic access to the room to replace batteries. Some listening devices can store information digitally and transmit it to a listening post at a predetermined time. Tiny microphones may be coupled with miniature amplifiers that filter out background noise.
- **Transmission Link:** The electrical impulses created by the pickup device must somehow be transmitted to a listening post. This may be done by a radio frequency transmission or by wire. Available wires might include the active telephone line, unused telephone or electrical wire, or ungrounded electrical conduits. Transmitters may be linked to an existing power source or be battery operated. The transmitter may operate continuously or, in more sophisticated operations, be remotely activated.
- **Listening Post:** This is a secure area where the signals can be monitored, recorded, or retransmitted to another area for processing. The listening post may be as close as the next room or as far as several blocks. Voice-activated equipment is available to record only when activity is present. A recorder can record up to 12 hours of conversation between tape changes.

Eavesdropping equipment varies greatly in level of sophistication. Many off-the-shelf spy shop devices are generally low-cost consumer electronic devices that have been modified for covert surveillance. They are easy to use against unsuspecting targets but can be detected by elementary electronic countermeasures. Devices produced for law enforcement and industrial espionage are more expensive, more sophisticated, and more difficult to find during a technical security countermeasures (TSCM) inspection. Devices designed and built for intelligence services are still more expensive and very difficult to find.

Some of the more sophisticated bugs have a "burst" transmission. A device about the size of a fingernail can record several hours of ordinary conversation and then transmit it to a remote receiver in a burst that lasts only two seconds. An hour of speech can be stored on a single chip. This is a passive system that records information but emits signals only when interrogated.[4] This makes detection very difficult. Of course, some countermeasures systems are designed to try to activate such systems so they can be detected.

Some eavesdropping operations, as discussed below, don't require anything at all to be planted in the target room. The eavesdropping can be done without ever having direct physical access to the target area. Such operations exploit weaknesses in the telephone system or computer system already in the target room, or they use a laser beam aimed at the target room.[4]

Eavesdropping in Office or Home

The type of bug installed in a home or office setting depends in part upon the length of time and the circumstances, if any, under which the installer has physical access to the site.

A visitor seated in front of your desk may bend down to pick up a dropped pen, using the few seconds when his hand is out of your sight to stick a bug under his chair or under your desk. Or he may "forget" and leave behind a workable pen that has a concealed microphone and transmitter. Any gift intended to be kept on your desk or elsewhere in the open in your office is a potential concealment device for a bug.

If the eavesdropper can gain a period of unsupervised access to your office or home, it is possible to install more sophisticated devices that are more difficult to detect. That is why physical security measures to protect the office space from intruders or other unauthorized persons are so important. Common hiding spots when time is available to plant a device include electrical outlets in the wall, furniture, lamps, ceiling light fixtures, pictures on the wall, books on your bookshelf, etc.

More than half of all eavesdropping attacks on U.S. offices, both foreign and domestic, have exploited the common telephone.[5] Telephones offer a variety of eavesdropping options, as the telephone instrument has electrical power, a built-in microphone, a speaker that can serve dual purposes, and ample room for hiding bugs or taps.

The time it takes to install a bug in your telephone is measured in seconds, not minutes. One type of telephone bug transmits all your telephone conversations to a nearby listening post. Picking up your telephone to make or receive a call triggers a recorder that can be placed in the trunk of a car parked up to four blocks away. When you hang up, the recorder is turned off automatically.

Another type of telephone bug will pick up conversations in the room and transmit them down the telephone line while your telephone remains on the hook. The eavesdropper can monitor your room conversations from another telephone anywhere in the world. Such telephone bugs are usually easy to detect by a professional countermeasures technician who knows what to look for.

With some of today's computerized phone systems, it is possible to manipulate a telephone electronically without ever having direct, physical access to the telephone instrument. Signals can be sent down the telephone line to turn the handset into a microphone that picks up and transmits conversations in the room even when the handset is hung up. This risk can be greatly reduced by the selection of an appropriate telephone system and implementation of available technical security countermeasures.

Computers are similar to telephones, in that they have the essential parts for a sophisticated surveillance system—a microphone and a means of communicating information outside the area in which they are located. Computers are vulnerable to several types of eavesdropping operations. For example, a bug in your keyboard could transmit every keystroke so that everything you write can be reproduced.

Standard computers emit faint electromagnetic radiation that a very sophisticated eavesdropper can use to reconstruct the contents of the computer screen. These signals can carry a distance of several hundred feet, and even further if exposed cables or telephone lines act as inadvertent antennas. Security measures and shielding are available to reduce the risk of such eavesdropping. It is possible to buy TEMPEST-protected computers that block the unintended radiation.

Eavesdropping in Public Places

Even public areas are not immune to technical surveillance. Whenever your presence in a public area is known or predictable in advance, an adversary or competitor has time to plan the best way to exploit that knowledge.

One Western European intelligence service is known to bug selected first class seats of its national airline. This picks up conversations among U.S. government officials or business executives traveling together for negotiations in that Western European country.

Outdoors in a park, in a hotel lobby, or while sitting around a hotel swimming pool, conversations may be monitored with a shotgun microphone. This is a directional microphone (parabolic reflector) that may be concealed in a sleeve or a folded newspaper and aimed at the target. Clarity of the recording may be improved by programs that cancel out extraneous noise and that employ neural net analysis to learn the target's speech patterns.

Individuals who habitually frequent the same restaurant or café and hold sensitive conversations over lunch or dinner are also vulnerable, especially if they usually sit at the same table or the restaurant manager cooperates with the eavesdropper. A short-term bug can simply be attached to the underside of the table. Longer term, one could build the bug into the table or into a vase or other item on the table. Although probably very rare, at least one highly-competitive, high-class restaurant is known to have bugged its own tables to obtain unfiltered feedback on customer reactions to the service and food.

References

1. Granite Island Group web site, www.tscm.com/whatistscm.html, citing a report by the Department of State, Bureau of Intelligence and Research, March 1997.

2. Great Southern Security web site, www.greatsouthernsecurity.com.

3. Steve Casimiro, "The Spying Game Moves into the U.S. Workplace," *Fortune*, March 30, 1998, p. 152. Lynn Fischer, "Technical Security: What is It? And Why Do We Need to Know about It," Security Awareness Bulletin, No. 2-96. Department of Defense Security Institute, August 1996.

4. Srikuman S. Rao, "Executive Secrets," *Forbes*, 1999.

5. Telephone Security Group, *National Telecommunications Security Working Group Information Series: Executive Overview*, January 1996.

Chapter 38

The Lazy Person's Guide to Wireless Network Security

How many of us would allow complete strangers to walk into our homes and listen to our daily conversations? No one that I know would. Ditto for the office, and especially any office where the information we discuss is vital to our success or survival. Let's try a third tack. How many network administrators do you know who would allow a complete stranger to walk into their wiring closet and plug a laptop to their company's network?

I doubt that anyone would, but the virtual equivalent is probably happening across America at this very moment. These strangers aren't physically plugging into networks, though. They are attaching to networks using wireless network technology, which grants the same level of access afforded by a physical connection. In this article, we'll look at some of the ups and downs of wireless network security, Wired Equivalent Privacy (WEP), work by various groups on wireless security issues, and some advice for securing your wireless networks.

Wireless Revisited

I first wrote about 802.11 wireless networking in the Winter 2000 issue of *CHIPS*. I believed then, and I still believe now, that we will move toward wireless systems over the next 10 years. Before we launch into our wireless security discussion, though, let's briefly review what a

---

"The Lazy Person's Guide to Wireless Network Security," by Major Dale Long, USAF (Ret.), Department of the Navy, http://www.norfolk.navy.mil/chips/archives/02_winter/index2_files/dalelongwinter.htm. Winter 2002.

wireless network is. Wireless Ethernet networks are built using radio waves. The Institute of Electrical and Electronics Engineers (IEEE) 802.11 standard defines the physical layer and media access control (MAC) layer for wireless local area networks (LANs). As with our wireless telephone networks, the basic building block of the 802.11 architecture is the cell, also known as the Basic Service Set (BSS). A BSS typically contains one or more wireless stations and a central base station. Base stations are the access points to the network and may be either fixed or mobile.

All the base stations in a particular wireless network communicate with each other using the IEEE 802.11 wireless MAC protocol. Multiple base stations may also be connected together using wired Ethernet or another wireless channel to form a distribution system that appears as a single 802 network in much the same way that a bridged, wired (IEEE 802.3) Ethernet network appears as a single network.

Like cellular telephone service, the cells in a wireless network overlap to provide coverage over an area larger than that covered by an individual cell. If a mobile user moves from one cell to another, the base stations should *hand-off* the user from one cell to another.

You can also get IEEE 802.11 stations together to form an ad hoc network with no central control and no connections to the outside world. The workstations form into a network simply because they happen to find they are in proximity (within the broadcast range) of other mobile devices that communicate in the same way even though there's no pre-existing network infrastructure (e.g., a pre-existing 802.11 BSS with an access point) in the area.

There has been explosive growth in the deployment of 802.11b networks over the last year. Much of the appeal of 802.11b networks can probably be credited to the Wireless Ethernet Compatibility Alliance (WECA), which developed the wireless fidelity (WI-FI) interoperability standard. Commercial products that bear the WI-FI logo must pass a suite of basic interoperability tests. When people plug into a WI-FI certified access point, it should work with any other WI-FI certified technology. Other growth factors are low cost and ease of installation. With close to 100 vendors offering the technology, prices have plummeted to under $100 for notebook cards and as low as $150 for access points. Physical deployment is extremely simple. All you have to do to install an access point out is take it out of the box, plug it into your wired Ethernet segment and turn it on. The combination of low cost and simplicity are powerful attractions.

Unfortunately for organizations with extensive wired networks, adding wireless access points to their network can subvert their entire security system because they are inside their network perimeter behind their firewall. WECA's goal for the WI-FI standard was interoperability and ease of use, not security. The 802.11b standard does include a provision for encryption called WEP (Wired Equivalent Privacy). Depending on the manufacturer and the model of the network interface card (NIC) and access point, there are two levels of WEP commonly available—one is based on a 40-bit encryption key and 24-bit initialization vector (IV). It is also called 64-bit encryption and is generally considered insecure. The other is a 104-bit key plus the 24-bit IV—also called 128 bit encryption. Unfortunately, even the 128-bit encryption, while stronger, is no longer considered completely secure.

Wireless Monitoring

There has been a lot of buzz in the computer and technology press over the last year about the basic insecurity of WEP. I got a first-hand exposure to this during my latest visit to a friend's house. My wife and I arrived for our monthly dinner with them to hear some joyful news: they're "expecting."

The fun began when they offered to show us the new nursery. It was a wonderful little room, full of bright colors, beautiful new baby furniture, and a Linux super-computing cluster. Yes, he had taken a simple concept like the baby monitor and once again elevated it to a project rivaling ballistic missile defense. The crib, changing table, mobiles, and stuffed toys were wired *to the gills* with networked sensors monitored and controlled by a computing system better than anything owned by Johns Hopkins University or NORAD [North American Aerospace Defense Command]. It could monitor heart rate, blood pressure, brain wave activity, blood sugar and a host of other activities too numerous, trivial, or disgusting to mention.

And, just to show off, he demonstrated that it could monitor a fly walking on the wall above the crib while simultaneously winning eight straight games of Internet speed chess against the British International Grandmaster, Nigel Short. At that moment, I remembered something odd I had seen on the way in. There were at least six vans parked on the street, each with at least two or three large antennas pointed toward their house. When I asked if they knew about them, he grinned. "Oh yeah," he said, "those are a few friends my wife invited over to help check security on our baby monitoring system."

She has some interesting friends, considering the vans included people from such diverse groups as the Federal Bureau of Investigation, AT&T, and the Central Intelligence Agency. When the pizza delivery guy arrived with dinner, the vans emptied and we all sat down for pizza and a discussion of wireless security. One of the vans wasn't there by invitation, though. It turned out to be a group from the SETI (Search for Extraterrestrial Intelligence) Institute who thought they'd finally found intelligent extraterrestrial life. They were very disappointed when my friend told them they'd been listening to a housefly breathe. As the evening unfolded, I got a real education on the trials and tribulations of security in the wireless networking world.

Tools of the Trade

There are apparently several ways to uncover patterns in packets of information passing over wireless LANs. These patterns can be used to figure out the WEP encryption key, which is the number used to scramble the data being transmitted. Once the key is recovered, it can be used to decrypt the messages.

A first tool in the wireless hackers' arsenal might be a wireless network scanner. One wireless scanner can allegedly discover WEP keys through passive monitoring. According to information located on its website, it can determine WEP keys in less than a second after listening to 100MB–1GB of traffic. And since many implementations of WEP are based on static keys that do not change over time, you can eventually sift out whatever data you need to crack the key—if you listen long enough.

Another wireless network scanner "sniffs" for wireless networks. When it identifies an 802.11b signal, it can paint a very accurate picture of the entire wireless implementation simply by "war driving" around the perimeter and locating the access points. Remember the movie *War Games* where the kid set his computer to dial every phone number in the area in sequence until he hit one associated with a computer? War driving is stocking up on good antennas and driving around randomly discovering wireless access points.

Many people assume that the 802.11b signals only travel a relatively short distance—maybe a hundred feet or so. They actually travel much farther, but are too weak to be detected by the tiny antennas in laptop cards. But with an external gain antenna, 802.11b signals can be detected at a much greater distance.

Locking Virtual Doors

Wireless network security is much like the physical security at the entrance of a building—anyone with enough time, interest and resources is going to be able to gain access. The trick is limiting the damage so the only things they see are what you want them to. First and foremost, we must treat wireless networks as publicly accessible at all times. We cannot assume that wireless traffic, in any media, is private and secure simply because the signal is always out there for someone to intercept.

Second, always enable WEP. Yes, WEP isn't considered totally secure at this point, but at least it's a first hurdle for people to cross. It's also free, so it costs you nothing to employ. Third, always change the default SSID (Service Set Identifier) of your system components. Don't, however, change the SSID to reflect your main names, divisions, missions or products. That can just make you a more interesting target when someone sees you have a server named after jet fighters or submarines. If your naming is interesting enough, it may attract hackers who are willing to put in the additional effort to break your WEP encryption keys. Probably the best way to deal with SSID problems is to disable "broadcast SSID" on your systems. By disabling that feature, the SSID configured in the client must match the SSID of the access point. You can configure your client systems to allow access—the cracker is on his own.

Fourth, change the default password on your access point or wireless router. Any good cracker will know the manufacturers' default passwords and will try them first. Since wireless network scanner programs identify the manufacturer based on the MAC address, it doesn't take much work to figure out what type of device it is even if you do change the SSID.

Finally, periodically survey your site using a tool like a wireless network scanner to see if any rogue access points pop up. It's not hard for some well-intentioned soul to go buy a couple of wireless cards and an access point and plug them into your wireless network. All of your best efforts at security could be wasted if a rogue access point is plugged into your network behind your firewall. Also, take a laptop equipped with the wireless network scanner and an external antenna outside your perimeter and check out what someone in your parking lot might "see." You'll be surprised how far the signal radiates. You might only connect at one to two MBps (megabytes per second), but it's still a potential security breach.

An Alternative to WEP

Since WEP has been written off as the principal source of security for wireless networking, various groups have been searching for alternatives. The NASA (National Aeronautics and Space Administration) seems to have found, at least for now, a working solution. The network security group in the NASA Advanced Supercomputing (NAS) Division at Ames Research Center believes that WEP provides no substantial security protection for the following reasons, some of which we have already examined:

1. Wireless card hardware addresses cannot be trusted as tools to identify a user;
2. The signal coverage perimeter cannot be easily limited to conform to an organization's physical control perimeter;
3. WEP encryption of data sent between a laptop and an access point can be cracked, regardless of key length;
4. Well documented cases have shown that deriving a WEP encryption key from hacked ciphertext and decrypting WEP traffic can be done without ever needing to derive the key.

In their implementation of the Wireless Firewall Gateway (WFG), NAS disabled all 802.11b network security features. Instead, all the services reached via the wireless network provide their own authentication. The WFG acts as a router between the wireless and external networks. It can dynamically change firewall filters as users authenticate themselves for authorized access. It is also responsible for handing out IP addresses to users, running a Web site in which users can authenticate, and maintain a recorded account of who is on the network and when. If you're interested in this approach, I encourage you to read the NAS white paper located at http://www.nas.nasa.gov/Groups/Networks/Projects/Wireless/index.html. They use OpenBSD UNIX and other open source networking products, so there may not yet be equivalent tools for other network operating systems. However, if it works well enough on OpenBSD, someone will eventually figure out how to port it to their own preferred brand.

Final Thoughts

As with any segment of technology, there are people trying to secure systems and people trying to crack that security. Be prepared, but not paranoid. We depend on our networks more and more every day, adding more eggs to that particular basket. The best advice I've found on this subject comes from Mark Twain, who wrote:

Behold the fool saith, "Put not all thine eggs in the one basket"—which is but a manner of saying, "Scatter your money and your attention;" but the wise man saith, "Put all your eggs in the one basket and—WATCH THAT BASKET."

Happy Networking!

—by Major Dale Long, USAF (Ret.)

Long is a retired Air Force communications and information officer.

Chapter 39

New Attack Intercepts Wireless Net Messages

It's the stuff of *Popular Science*. A group of security researchers has discovered a simple attack that enables them to intercept Internet traffic moving over a wireless network using gear that can be picked up at any electronics store and an easily downloadable piece of freeware.

The attack, accomplished by @Stake Inc., a security consulting company in Cambridge, Mass., affects a popular consumer version of Research In Motion Ltd.'s BlackBerry devices as well as a variety of handhelds that send unencrypted transmissions over networks such as Mobitex.

By design, the Mobitex specification, like other wireless standards such as Global System for Mobile Communications and General Packet Radio Service, sends packets in unencrypted form. The network, which handles data transmissions only, has been in operation since 1986 and has a large base of installed devices, with customers using it for everything from point-of-sale verification to e-mail.

"The attack is fairly simple," said Joe Grand, one of the researchers who perfected the technique. "The problem is, this isn't a bug. It's part of the spec that data is transmitted in the clear, just like it's part of the spec that Internet data is transmitted in the clear. The risk

---

"New Attack Intercepts Wireless Net Messages," Ziff Davis Media, Inc., http://www.eweek.comprint_article/0,3668,a=23806,00.asp. Reprinted from eWeek, March 11, 2002, with permission.

depends on who is using the network and when and what data they're sending."

Using a scanner with a digital output, an antenna and freely downloadable software, the researchers were able to intercept traffic destined for BlackBerry Internet Edition devices. And, because the packets aren't encrypted, the attackers can read the messages they intercept without further work.

The Internet Edition handhelds are sold mainly through co-branding relationships with ISPs such as AOL Time Warner Inc.'s America Online service, EarthLink Inc. and Yahoo Inc.

Executives at RIM said they don't see the attack as a problem because they have never touted the Internet Edition devices as being secure.

"Internet traffic isn't supposed to be secure," said Jim Balsillie, chairman and co-CEO of RIM. "It's kind of like a company making beer and cola and someone saying that there's alcohol in the company's drinks when the children are drinking cola."

However, the attack serves as a reminder to users that e-mail and other Internet traffic is open to snooping and is inherently insecure.

"I always figure that anything that's sent via e-mail can be read by at least hundreds of people which have either legitimate or compromised access to systems sitting between me and my recipient; this just adds another potential access point," said Christopher Bell, chief technology officer of People2People Group, a relationship services company in Boston, and a user of the BlackBerry Internet Edition. "I am disappointed that they didn't make at least a modest attempt to obscure the content."

Balsillie said the messages are only as secure as the networks of the ISPs that relay them, none of which provide encrypted e-mail.

Chris Darby, CEO of @Stake, said RIM has done a thorough job including security in its other devices, which use a server that sits behind corporate firewalls.

"RIM is incredibly progressive about the way they're addressing security in their Enterprise Edition," Darby said.

The attack also applies to other devices on the Mobitex network, many of which are proprietary solutions developed for in-house corporate uses.

This attack does not work on the BlackBerry Enterprise Edition, which uses Triple Data Encryption Standard encryption in addition to other security features, @Stake officials said.

"Typically, Mobitex operators will advise customers that they should choose the security scheme that fits their particular needs," said Jack Barse, executive director of the Mobitex Operators Association, based in Bethesda, Md. "It was a conscious decision not to put network-level security in because customers have said that they don't want the overhead associated with security if they're just doing things like instant messages. Customers can absolutely add on their own encryption to whatever application they're using [the network] for. And we encourage that."

—by Dennis Fisher and Carmen Nobel

Chapter 40

Aircracked! Wireless Hacking and "Drive-by" Cracking

It may not be illegal. There are free tools that help you do it. It's wireless hacking and it's highly dangerous to your information resources.

Imagine you're invisible. You walk along a downtown New York City (NYC) street deep in the financial district. You can even wear your pajamas. You turn into a gleaming skyscraper, walk past the security guard and get in the elevator. You get off on the top floor with a serious-looking executive and walk into the corner office with him. There, you settle into one of the comfortable chairs (forgoing the Cuban cigars arrayed on the coffee table). Then, you wait. After a while, the executive calls in another serious-looking executive. They confer on the impending acquisition of BigBank by BiggerBank and congratulate each other on the coup they are about to score. You get up, walk out the door, leave the building and call your broker. "Buy shares of BigBank," you instruct him. You go home, get out of your pajamas and enjoy an afternoon sailing on your yacht.

Sound far-fetched? Let's push it further. You drive out to the suburbs, walk right into the bedroom of BiggerBank's CEO and hang around listening to him talk to various friends and colleagues about the next big transaction—BiggerBetterBank.

"Aircracked!," SC On-Line, SC Magazine, 161 Worcester Road, Suite 201, Framingham, MA 01701, http://www.scmagazine.com/scmagazine/sc-online/2002/article/39/article.html. September 2002. Copyright © West Coast Publishing. Reprinted with permission.

Congratulations! You're part of the new wireless hacker, drive-by cracker crowd. And yes, you can wear your pajamas if you insist. Armed with tools bearing names such as *NetStumbler* and *AirSnort*, wireless intruders are harnessing high-powered receivers and transmitters to break into corporate wireless networks from as far away as ten miles. Throw in some simple GPS software and you can instantly map the location of each unprotected wireless access point you find. And there are thousands of these.

Not since the invasion of the Macintosh has a new technology invaded the corporate IT infrastructure with such ferocity. Wireless access points spring up like weeds under desks, in drawers, in ceilings and closets. At least half the time they are left unprotected. Why? Friends, teenage sons or the neighbor techie accompany the well-meaning business person—who is sick of having to find plugs in conference rooms—perhaps on a weekend day, to install a wireless access point somewhere near the office Ethernet jack. The following Monday the executive walks around the office with newly untethered freedom.

Accompanying him, of course, are the invisible aircrackers.

How bad is the problem? On a recent leisurely drive through, yes, the financial district in downtown Boston, Crossbeam engineers discovered several hundred unprotected wireless access points that gave them access to major corporate IT resources in banks, financial services firms, law firms, consultancies—you name it.

Now, these engineers weren't using anything more sophisticated than an external car antenna, a Lucent wireless card and *NetStumbler*. Fortunately, they were just geeks having fun with no intention of looking at proprietary content.

Consider, though, the case of an intrusion suffered several months ago by a major Fortune 500 company. The intrusion was executed from twelve miles away and the intruder was able to hop onto the corporate network and penetrate deep into the company's IT infrastructure. Did the intruder take anything? Who knows? However, the IT security staff detected an unknown machine on the network, suspected an unprotected wireless access point and via a painstaking search of every floor of the breached building found not one but several of the rogue access points and shut them down. Did this stop their worries? Of course not. For one, they knew that new unauthorized access points would spring up. Even worse, they knew that protected access points are crackable.

Here's the depressing news—the wireless encryption protocol (WEP) is not secure. It turns out that while the encryption algorithms at the root of WEP are supposed to generate random keys, in practice they don't do this. If you wait long enough, you will see keys repeat and this is enough for the crypto-crackers to get their hooks into the pattern and eventually break the keys. By the way, you don't have to be a rocket scientist to break the key—a free, publicly available PERL script called *WEPCrack* will do it for you.

Here's the really interesting part: if you download a file from a corporate network via a wireless connection, have you committed a crime? Well, the jury's out. What? you say, How can this be?—you, Mr. or Ms. Aircracker, are stealing intellectual property or financial records that belong to someone else! Ah yes, but here's the catch. Because the wireless access point is *broadcasting* its service availability and because the company does not charge for this "service" (just like an FM station), it may not be considered theft of service. Basically, companies could be construed as "inviting" hackers and crackers into their network! Just like talk radio—only much more interesting.

So, given this technological and legal chaos, what can you do to protect your corporate information assets? Obviously, you will need to create extraordinary awareness within your company that this is a huge threat. However, knowing that the human factor is the least dependable, there are two technology solutions that will dramatically reduce the likelihood of a breach: one protects your building from external eavesdropping, the other forces authenticated access to the network.

For the external eavesdropping solution let's go back to our Fortune 500 company. Through a stroke of good luck, the company happened to be in the process of moving much of its IT infrastructure into a new building. However, the building was not nearly complete at the time of the wireless intrusion. Our smart network security engineers alerted the building project managers and together they found a vendor who could supply shielded glass for the windows in the building. By testing various flavors of glass, they picked one that effectively blocked access point broadcast advertisements beyond the walls of the building.

The tougher part of the solution was to force all wireless client access through an authentication mechanism that was not WEP. The solution they found was to place a lightweight virtual private network (VPN) server running *Check Point VPN-1* between every access point and its network connection. This meant that in order to jump onto the network over a wireless connection, an end user would first have to type in a user name and password and then create an IPsec tunnel to access the network. Because the username and password are encrypted and because IPsec uses a much stronger encryption scheme than WEP, the company's network is now much more protected than it was.

Does it solve the problem completely? Of course not, since the next unprotected, unauthorized access point that springs up creates an instant vulnerability. Although the company is not there yet, they are now talking about creating an "interior VPN" mesh so that all access to the network is authenticated, whether wireless or not. This would mean placing VPN termination points at least in front of major corporate information assets in data centers and eventually extending out to the furthest leaves in the network.

From an administrative point of view, this requires a VPN solution that has extraordinary global VPN management for add/change/deletion of rules and policies as well as global software updates for revisions and patches. The good news is that products like Check Point's are mature enough to have developed good management tools. The difficulty will be creating the project to put such an infrastructure in place. For while the cost of intrusions is potentially very large in terms of both dollars and lost reputation, the cost of large network upgrades is also high. However, even here, by starting with data center assets, the greatest concentration of corporate information will be protected.

Ultimately, there is a larger issue, of which wireless vulnerabilities are just one element. A large percentage of security intrusions originate from inside corporate networks—that is, their own employees. Consequently, the "interior VPN" by itself is not enough. This is why some companies are supplying a new level of integrated, best-of-breed security devices that provide "decontamination" nodes where traffic can not only be VPN'd but also checked for intrusion signatures, viruses and malicious web code before entering or exiting a VPN tunnel—all at wire speed.

Whatever your decision as an IT or network manager might be, it would be highly instructive to go crack your own network and decide for yourself how vulnerable your company is and what the cost of a breach might be. Then, ask your executive management how they feel about the magnitude and implications of a breach. You might just find new budget dollars springing up like weeds in closets, drawers and all kinds of unexpected places!

—by Throop Wilder

Throop Wilder is co-founder and vice president of marketing for Crossbeam Systems, Inc.

Chapter 41

Roving Wiretaps: Explanation and Analysis

How the Bill Would Change Foreign Intelligence Surveillance Act (FISA) Surveillance

General Comments

The Anti-Terrorism Act (ATA) is defended on the basis of needing to meet the threat of terrorism. But ATA's expansion of FISA powers is not limited to terrorism cases. It must not be forgotten that a major reason for FISA was a well-documented record of executive branch abuse of national security surveillance powers.

Second, the government essentially claims that FISA is more restrictive than other surveillance statutes, and that FISA should "more closely track" such other laws. This argument strikes EFF as disingenuous, because the government's special FISA powers are sui generis, intended only for gathering intelligence about foreign powers for counterintelligence purposes.

Third, a recurring theme of ATA is the elimination of the requirement that FISA surveillance be limited to "agents of a foreign power." This requirement protects U.S. persons (citizens and permanent resident aliens), to some extent, against FISA surveillance. Removing this

---

"EFF Analysis of The Anti-Terrorism Act of 2001 (ATA)," Electronic Frontier Foundation, http://www.eff.org/Censorship/Terrorism_militias/20010927_eff_ata_analysis.html, September 27, 2001. The full version of this document, including references, is available on the Electronic Frontier Foundation website at www.eff.org.

requirement is likely to increase FISA surveillance of U.S. persons. Non-U.S. persons, of course, are also entitled to Fourth Amendment protection.

ATA Sec. 151

Under current law, FISA authorizes:

- electronic surveillance of "agents of a foreign power" for up to 90 days, and of "foreign powers" for up to 1 year. 50 United States Code (U.S.C.) § 1805(e)(1);
- physical searches of "agents of a foreign power" for up to 45 days, and of "foreign powers" for up to 1 year. 50 U.S.C.§ 1824(d)(1).

This section would extend the duration of a FISA order to up to one year for "agents of a foreign power" under § 1801(b)(1)(A), which applies to non-U.S. persons, i.e., aliens in the United States who are not permanent residents.

This is a bad idea, because aliens in the United States are entitled to the protection of the Fourth Amendment. Moreover, this extension would apply to surveillance of their home phones and computers, because their offices can already be bugged or wiretapped for a year.

By comparison, Title III provides that ordinary wiretaps may not last more than 30 days, and each successive extension is equally limited. 18 U.S.C. § 2518(4).

Also, it should be remembered that the initial order is based only a showing of probable cause. The point of the 45- and 90-day limits is to require the government to come back and demonstrate, armed with the experience of the initial search or surveillance, that the target truly is an "agent of a foreign power." Extending the duration of initial order means that persons who are not "agents of a foreign power" are subjected to unjustified search and surveillance for a much longer period.

ATA Sec. 152

Under current law, FISA does not authorize so-called "roving wiretaps." Roving wiretaps are unlike conventional wiretaps in that they allow law enforcement officials to follow the suspect from one location to the next, without having to seek court authorization to wiretap each location's telephone line or other communication channel. In short, the government may wiretap any telephone that the target uses or is known to use.

Roving wiretaps pose serious problems under the Fourth Amendment, which requires that any search warrant "particularly describ[e] the place to be searched, and the person and things to be seized." The particularity requirement carefully tailors the scope of search to its justification. An insufficiently particular warrant may constitute an unconstitutional "general warrant" like those used by the British against American colonists—a prime concern of the Framers.

In *Andresen*, for instance, a real estate attorney was suspected of fraud, and a search warrant was executed on his office. The officers seized substantial information beyond what the warrant specified because of a "catch-all" phrase in the warrant. The Supreme Court held that the "catch-all" phrase made the search general.

Equally important, when the Court definitively ruled on the constitutionality of eavesdropping, it made clear that unless the particularity requirement was observed, an officer would have a "roving commission to 'seize' any and all conversations." Merely naming the person, the Court said, does not "particularly describ[e] the communications, conversations, or discussions to be seized."

This section would expand FISA to include "roving wiretap" authority. Current law requires court-"specified" third parties (like common carriers and ISPs) to provide assistance necessary to accomplish the surveillance. The proposed change would extend that obligation to unnamed and unspecified third parties. According to the Justice Department, "the FBI could simply present the newly discovered carrier, landlord, custodian, or other person with a generic order issued by the Court, and could then effect FISA coverage as soon as technically feasible."

In practical terms, roving wiretaps pose a greater danger to personal privacy than ordinary wiretaps for two reasons. First, the issuing court in effect gives the government a blank check to do surveillance, because it does not approve particular wiretaps. There simply will not be a showing of probable cause for each wiretap.

Second, by extending surveillance to many more communication channels, the number of potentially innocent conversations of innocent persons increases. "[O]nce a roving intercept order is issued, there is no express limitation on the number of places in which the government can install listening devices or telephones it can tap, and the decision in each instance [is] an executive rather than a judicial one."

The impact of this amendment would be especially great for communication facilities used by the general public, from public payphones to computers in public libraries. Upon the suspicion that a FISA target might use such a facility, the FBI could monitor all communications transmitted at the facility, and the recipient of the assistance order could not disclose that monitoring is occurring.

ATA Sec. 153

Under current law, FISA surveillance may only be used when foreign-intelligence information gathering is "the" sole or "primary" purpose.

This section would permit FISA surveillance even when the main purpose is to investigate a crime, which destroys the existing balance between counterintelligence and law enforcement surveillance.

When the FISA court reviews an application for FISA surveillance, its job is merely to assure that all the necessary certifications—including the representation that the primary purpose is to gather foreign intelligence information—are present and not clearly erroneous. Thus, when FISA surveillance is challenged in a criminal proceeding, courts have said that it is "not the function of" the courts "to 'second-guess' the certifications."

Given that FISA contains far fewer procedural protections than Title III, and that FISA orders are issued and implemented in great secrecy, this is an enormous change in the law.

ATA Sec. 154

Under current law, information obtained from non-FISA criminal investigations like federal grand jury investigations may only be disclosed under stringent procedural safeguards.

This section would allow "foreign intelligence information" gathered in such investigations to be shared with federal law enforcement, the intelligence and defense community, and immigration authorities.

The secrecy of grand jury investigations under current law is partly due to the sweeping powers of grand juries to compel the disclosure of information via subpoenas without judicial oversight. The Supreme Court has emphasized that the grand jury "is a grand inquest, a body with powers of investigation and inquisition, the scope of whose inquiries are not to be limited narrowly by questions of propriety or forecasts of the probable result of the investigation, or by doubts whether any particular individual will be found properly subject to an accusation of crime."

The grand jury can call anyone to testify before it based on a prosecutor's speculation about possible criminality, and they can be asked to bring documents and other tangible things. Any aspect of a person's life that might shed some light on criminality by someone is within the scope of a grand jury investigation.

In contrast, the government cannot get an ordinary search warrant to obtain evidence unless it has probable cause.

Thus, this section also upsets the balance between counterintelligence and law enforcement surveillance: it creates an incentive to use grand jury and other investigations as a tool for foreign intelligence collection.

ATA Sec. 155

Note that the FISA definition of pen/trap devices refers to the definition in the main pen/trap statute, which would be expanded under ATA Sec. 101. (see EFF's analysis of the Combating Terrorism Act [CTA] for a discussion of pen/trap devices).

Under current law, pen/trap devices may not be used under FISA unless the government shows that it will be placed on a communications device that has been or will be used in communications with "an agent of a foreign power."§ 1842(c)(3).

This section eliminates that requirement, allowing FISA pen/trap orders to be used on a government certification that it is likely to obtain information relevant to an ongoing foreign intelligence or international terrorism investigation.

Thus, FISA pen/trap surveillance could be used against any person, not only agents of foreign powers. This eliminates two key protections for U.S. persons under FISA (as opposed to non-resident aliens). First, there would be no showing that the U.S. person was involved in some kind of criminality, which is needed for a U.S. person to be an agent of a foreign power.§ 1801(b)(2). Second, it would evade the FISA constraint that a U.S. person cannot be deemed an "agent of a foreign power" solely on the basis of First Amendment activities.§ 1805(a)(3)(A).

ATA Sec. 156

Under current law, a limited set of records—those of common carriers, public accommodation facilities, physical storage facilities, and vehicle rental facilities—can be obtained with a court order. 50 U.S.C.§ 1861-62.

This section gives the government the authority to obtain "any tangible things," including documents, via administrative subpoena, so long as they are relevant to a foreign intelligence or international terrorism investigation.

This section greatly expands the scope of the "business records" provision. Moreover, the use of subpoena power eliminates the possibility of judicial oversight, because a court order would be unnecessary.

The government's only argument for this substantial change is that "[t]he time and difficulty involved in getting such pleadings before the Court usually outweighs the importance of the business records sought." This strikes EFF as remarkably weak. If the records are not important, then the authority is unnecessary. If the records are, in fact, important, then it should be worth applying for a court order. Note also that the government's section-by-section analysis does not even mention that ATA would expand the scope of this section to business records in general.

ATA Sec. 157

Current law generally requires that government access under FISA to a variety of records (governed by the Fair Credit Reporting Act, the Financial Right to Privacy Act, the Encrypted Communications Privacy Act [ECPA]) is conditioned on a showing by "specific and articulable facts" that there is reason to believe that the entity, person or consumer is an "agent of a foreign power."

This section would eliminate this requirement and permit such access if the government certifies that the information is "relevant to an authorized foreign counterintelligence investigation."

Not only does this lessen the government's factual burden, it removes the protections for U.S. persons that accompany the "agent of a foreign power" requirement.

ATA Sec. 158

The Federal Education Rights and Privacy Act protects the privacy of various educational records.

This section would permit government access to such records if "any" federal employee designated by the Attorney General or Secretary of Education determines that the records can reasonably be expected to assist in investigating or preventing terrorism.

Note that because ATA Sec. 309 defines many computer crimes unrelated to terrorism as "Federal terrorism offenses," this section would eliminate student privacy for all records that might relate to investigating such crimes. This represents a significant violation of privacy unrelated to the ATA's purported anti-terrorism justification.

—by Lee Tien

Lee Tien is a Senior Staff Attorney for the Electronic Frontier Foundation.

Chapter 42

Cell Phone Security: Taking Steps against Eavesdropping and Cloning

Cell phones are more vulnerable than regular phones due to two dangers: eavesdroppers can listen in on your calls, and thieves can bill their own calls to your account.

Eavesdropping: Anything you say on an analog cell phone can be easily overheard by someone using a scanner. Digital cell phone transmissions are scrambled for better protection, but eavesdroppers with the right equipment may be able to unscramble them.

The best protection? Be aware of what you discuss on your cell phone. Remember that it acts as a handheld broadcast station. Don't give out your credit card number or other sensitive or confidential information; don't say anything you wouldn't say on broadcast radio or TV.

Fraudulent billing: it is possible for thieves to intercept a cell phone signal and clone the phone's ID numbers (its Electronic Serial Number and Mobile Identification Number, or ESN/MIN). The result is the equivalent of a stolen calling card. Some simple countermeasures include:

- **Limit "roaming":** Review which phones have roaming enabled and limit these as much as practical. Roaming usually defeats

---

"Cell Phone Security," Cornell University, http://www.cit.cornell.edu/cellphone/security.html. Reprinted with permission from Cornell Information Technologies (CIT), Cornell University, Ithaca, NY.

the use of Personal Identification Numbers (PINs). Cloners prefer roaming phones for this reason and they target airport parking lots, airport access roads, and rural interstates. Roaming also makes it more difficult for some cellular carriers to use fraud-detection programs to monitor an account and shut it down when fraud is detected.

- **Turn the phone off.** Cell phones poll the cellular base station with the strongest signal every few seconds. This is how the system knows which base station to route calls through. However, this polling exposes the phone to interception and cloning.
- **Review all bills** and report every erroneous call to the service providers. There are two types of cloning:
  - Outright theft of the phone's ESN/MIN is most common. A bill will reflect hundreds, even thousands, of bogus calls.
  - The other type of cloning is called tumbling, where a cloned phone uses a different ESN/MIN for each call. A bill might have only one bogus call this month, none next month, but three calls the month after that. The phone has still been cloned and fraud is occurring.
- **Prefer hands-off vehicle-mounted phones to handhelds.** The boxes used to capture ESN/MIN have a limited range; cloners will follow an individual they know is using a phone. Recent news reports reflect the chances of an accident increase substantially if a driver is operating a vehicle and a cellular phone simultaneously.

Chapter 43

Detection and Prevention of Eavesdropping

Any indication that an adversary or competitor is using illegal means to collect information should alert you to the possibility, at least, that listening devices might be planted in your office or home. There are a number of specific warning signs that you may be the target of eavesdropping. Of course, if eavesdropping is done by a professional, and done correctly, you may not see any of these signs.

One of the most common indicators of eavesdropping is that other people seem to know something they shouldn't know. If you learn that an activity, plan, or meeting that should be secret is known to an adversary or competitor, you should ask yourself how they might have learned that.

An eavesdropper will often use some pretext to gain physical access to your office or home. It is easy for an outsider to gain access to many office buildings by impersonating a technician checking on such things as the air conditioning or heating. The only props needed are a workman's uniform, hard hat, clipboard with some forms, and a belt full of tools. If challenged, the imposter might threaten not to come back for three weeks because he is so busy. In one version of this technique, the eavesdropper actually causes a problem and then shows up unrequested to fix it. In other words, you must verify that anyone performing work in or around your office was actually requested and

---

"Detecting and Preventing Eavesdropping," from an undated website produced by the Counterintelligence Training Academy, Nonproliferation and National Security Institute, Department of Energy (DOE), http://www.smdc.army.mil/Intelligence/Security%20Guide/V3bugs/Detect.htm. Downloaded March 2002.

is authorized to do this work. If a worker shows up without being asked, this suggests an attempted eavesdropping operation and should be reported immediately to your security office. Even when the work is requested, outside service personnel entering rooms containing sensitive information should always be accompanied and monitored.

Gifts are another means of infiltrating a bug into a target office. Be a little suspicious if you receive from one of your contacts a gift of something that might normally be kept in your office—for example, a framed picture for the wall or any sort of electronic device. Electronic devices are especially suspicious as they provide an available power supply, have space for concealing a mike and transmitter, and it is often difficult to distinguish the bug from other electronic parts. Have any gift checked by a technical countermeasures specialist before keeping it in a room where sensitive discussions are held.

Unusual sounds can be a tip off that something is amiss. Strange sounds or volume changes on your phone line while you are talking can be caused by eavesdropping. However, they can also be caused by many other things and are relatively common, so this is not a significant indicator unless it happens repeatedly. On the other hand, if you ever hear sounds coming from your phone while it is hung up, this is significant and definitely should be investigated. If your television, radio, or other electrical appliance in a sensitive area experiences strange interference from some other electronic device, this should also be investigated if it happens repeatedly.

Illegal entry to your office or home to install an eavesdropping device sometimes leaves telltale signs, especially if done by an amateur. Evidence of improper entry with nothing being taken is suspicious. Installing an eavesdropping device sometimes involves moving ceiling tiles, electrical outlets, switches, light fixtures, or drilling a pinhole opening in the wall or ceiling of the target room (drilling in from the other side of the wall or ceiling). This can leave a small bit of debris, especially white dry-wall dust that should not be cleaned up. It should be reported to the security office.

In summary, protection against the installation of eavesdropping devices requires:

- Alert employees.
- Round the clock control over physical access by outsiders to the area to be protected.
- Continuous supervision/observation of all service personnel allowed into the area for repairs or to make alterations.

- Thorough inspection by a qualified technical countermeasures specialist of all new furnishings, decorations, or equipment brought into the area.

What to Do If You Suspect You Have Been Bugged

If you suspect you are bugged, do not discuss your suspicions with others unless they have a real need to know. Above all, do not discuss your suspicions in a room that might be bugged. Do not deviate from the normal pattern of conversation in the room. Advise your security officer promptly, but do not do it by phone. The bug may be in the telephone instrument. Do it in person, and discuss the problem in an area that you are confident is secure.

These security measures are important to ensure that the perpetrator does not become aware of your suspicions. A perpetrator who becomes aware you are suspicious will very likely take steps to make it more difficult to find the device. He may remove the device or switch it off remotely.

Never try to find a bug or wiretap yourself. What's the point? If you are suspicious enough to look, you already know you should not have any sensitive conversation in that room. If there is a bug there, do-it-yourself approaches probably will not find it. If you look and don't find it, that certainly shouldn't give you any sense of confidence that you can speak freely in that room. Don't be misled by what you see on television, in the movies, or in spy-shop catalogs. Detecting bugs is difficult even for the professionals who specialize in that work.

Technical Security Countermeasures

A Technical Security Countermeasures (TSCM) survey, also known as a "sweep," is a service provided by highly qualified personnel to detect the presence of technical surveillance devices and hazards and to identify technical security weaknesses that could facilitate a technical penetration of the surveyed facility. It consists of several parts.[1]

- An electronic search of the radio frequency (RF) spectrum to detect any unauthorized emanations from the area being examined.
- An electronically enhanced search of walls, ceilings, floors, furnishings, and accessories to look for clandestine microphones, recorders, or transmitters, both active and quiescent.

- A physical examination of interior and exterior areas such as the space above false ceilings and heating, air conditioning, plumbing, and ventilation systems to search for physical evidence of eavesdropping.
- Identification of physical security weaknesses that could be exploited by an eavesdropper to gain access to place technical surveillance equipment in the target area.

During the survey, TSCM team members may enter office areas where employees are working. Employees should be advised in writing, not orally, that a technical security inspection is being conducted and that they should not discuss it in the office before, during, or after the survey.

Reference

1. This section is based on an article by James Calhoun, “Clear the Air with TSCM.” Department of Defense Security Institute, *Security Awareness Bulletin 2-96*. The article appeared previously in *Security Management* magazine.

Chapter 44

Voice Mail, Answering Machine, and Fax Surveillance Vulnerabilities

Voice Mail

The remote access feature of voice mail makes it vulnerable to monitoring. You do not need to be in your office to receive your voice mail. You can call from home or any other location, dial a password to identify yourself, and hear your messages. The problem is that any other person calling the same number and using the same password can also retrieve your messages.

The password is usually easy to guess, because few people take the trouble to protect their voice mail with a unique password. They do not change the default password that comes with the system when it is installed. This is often the last four digits of the telephone number

---

Excerpted from "Voice Mail," from an undated website produced by the Counterintelligence Training Academy, Nonproliferation and National Security Institute, Department of Energy (DOE), http://www.smdc.army.mil/Intelligence/Security%20Guide/V2comint/Voice.htm, downloaded March 2002; "Answering Machines," from an undated website produced by the Counterintelligence Training Academy, Nonproliferation and National Security Institute, Department of Energy (DOE), http://www.smdc.army.mil/Intelligence/Security%20 Guide/V2comint/Answring.htm, downloaded March 2002; "Fax Machines," from an undated website produced by the Counterintelligence Training Academy, Nonproliferation and National Security Institute, Department of Energy (DOE), http://www.smdc.army.mil/Intelligence/Security%20Guide/V2comint/Fax.htm, downloaded March 2002; and "Telephones," from an undated website produced by the Counterintelligence Training Academy, Nonproliferation and National Security Institute, Department of Energy (DOE), http://www.smdc. army.mil/Intelligence/Security%20Guide/V2comint/Telephon.htm, downloaded March 2002.

or the employee's extension number followed by the pound sign. People who do change the password often use an easily guessed password such as their first or last name or date of birth.

Any current or former employee who knows the voice mail phone number and can guess your password can listen to your voice mail. In many cases, that will make no difference as your voice mail contains no information that requires protection. Here are a couple examples of cases where it did make a difference, however.

Michael Gallagher, a reporter for the Cincinnati Enquirer, stole voice mail messages from Chiquita Brands International, Inc., the banana company, and used them as the principal source for an 18-page series of articles that exposed the company's business practices. He allegedly stole thousands of voice mail messages with the help of three current or former Chiquita employees. The reporter pleaded guilty in September 1998 to two felony charges—unlawful interception of communications and unauthorized access to computer systems, and the newspaper agreed to pay Chiquita more than $10 million in damages.[1]

In another case, John Hebel, a disgruntled former employee of Standard Duplicating Machines Corp. (Standard) of Andover, MA, regularly broke into the voice mail system of his former employer as part of a scheme to make unauthorized use of the company's sales leads and confidential marketing information.[2] Hebel was a field sales manager who worked out of his home in Ballwin, MO. After being terminated by Standard, he went to work for a competitor as its Midwest Regional Manager.

Hebel developed a scheme to defraud his former employer by gaining unauthorized access to its voice mail system. By virtue of his prior employment at Standard, Hebel knew the telephone number for accessing Standard's voice mail system from remote locations. He knew that the "default" password for a particular voice mailbox would be the employee's telephone extension plus the pound sign, and that virtually no Standard employees had utilized unique passwords to protect their voice mail boxes. Hebel also knew which Standard executives and employees were likely to receive sales leads and other confidential marketing information in their individual voice mail boxes.

Over the course of a year, Hebel stole information from Standard's voice mail system on several hundred occasions. Standard eventually learned of Hebel's activity through an unsolicited phone call from a customer who had been solicited by Hebel after leaving a message on Standard's voice mail system. The FBI arrested Hebel for wire fraud, and he was sentenced in March 1997 to two years probation.

Answering Machines

Answering machines are the handy helper of many busy people. We are not always available to receive important telephone calls, so we choose is to install an answering machine. Although these machines are handy, they can raise security and privacy issues.

The principal vulnerability of answering machines is very similar to voice mail. The remote access feature makes it possible for unauthorized persons to dial your number and listen to your messages. Your only protection is a short code number that is pre-set in the factory and which very few buyers of answering machines bother to change. The factory-set code numbers for various models of answering machines are known to those who specialize in this type activity. Therefore, anyone who knows your phone number and the factory-set code number can gain access to your answering machine. Even if you have changed the pre-set code number, the new code number may not be difficult to break.

Many home answering machines also have a second vulnerability. They are equipped with security features that allow the owner to telephone home and listen for burglars or other activity in the home. In this case, the telephone instrument serves as a microphone listening for sounds. The designers and builders of the machines didn't intend for this feature to be used by anyone other than the owner, so they built in a security code similar to the code used for remote access to your messages. The feature works like this: you dial your home telephone number followed by the security code number. Before the telephone rings the answering machine picks up the call and allows the caller to listen for conversations or sounds within the house.[3]

Consider what might happen if an eavesdropper knew your code or could easily break it. Now consider that manufacturers make thousands of these machines with a factory-set security code. If this is not changed after purchase, as many people fail to do, anyone who knows the factory-set security code could easily call your machine and listen to any private conversations within earshot of the answering machine. Even if you have changed the factory-set security code, the code is susceptible to being broken.

Fax Machines

Fax machines have proliferated to almost every business and many homes. This method of communication is one of the most easily and successfully targeted. The means for intercepting fax signals are the same as for telephones, detailed below. Countermeasures for protecting the security of fax transmissions are the same as the countermeasures for protecting telephone conversations.

Stand-alone and PC-based fax and data intercept equipment is available that will covertly intercept and capture fax transmissions from any make or model machine, regardless of its "handshake" (the signal that two machines will emit to get synchronized to send the message). It will also decipher nonstandard fax protocols and perform automatic intercept and storage.[4]

Some fax machines have another vulnerability—remote access to stored fax images. For example, the hidden mailbox feature is used to reserve fax documents for privileged users. The privileged user can retrieve the document(s) by inputting a personal identification number (PIN) just like your automatic teller machine (ATM) cards. The invasion of one's fax could occur when an unauthorized user has access to your PIN. Like other code numbers and passwords, the PIN can often be guessed or broken.[3]

How Fax and Telephone Communications Are Transmitted—And Can Be Intercepted

Satellite Transmissions and Land-based Microwave

When you pick up a telephone to make a call or send a fax, you generally have no idea through what channel the call will be routed. Automatic switching equipment routes the call by land line, by land-based microwave relay towers, or via satellite depending on which method is available and most efficient at the time.

Most long distance calls travel at least part of the way via the airwaves—by satellite or between land-based microwave towers—and anything in the air can be intercepted. The technology for monitoring microwave and satellite communications used to be so expensive and complicated that it required a major government investment. Now, any reasonably well-financed group or individual can do it with readily available, off-the-shelf equipment.

The advent of digital communications has brought a large increase in the number of simultaneous digital voice or data transmissions over a single communications media. However, technological advances in high-speed computer search engines have kept up with this increase in volume. Therefore, it is still easy to sort through the billions of telephone calls and fax and data transmissions to identify targeted phone numbers or do key-word searches to pick out calls that mention watch-listed topics. Manpower requirements for processing the voluminous intercept material are greatly reduced by doing the initial screening by computer. Communications intercepted in the United States may be relayed back to another country for analysis and translation of significant messages.

Here's how intercepting telephone and fax communications works.[5]

Satellite

Let's suppose your signal has traveled by land line or land-based microwave to a long-haul switching station. The telephone company's computer searches for the most efficient path to send the signal and picks out a satellite connection. Your call is relayed to a ground station where it is transmitted by a transponder up to a satellite and then relayed down to a distant ground station. The call then goes via land line or land-based microwave to a switching station where it is unlinked from the other signals, passed over cable to the recipient's telephone, and converted back into voice or a fax message.

All this happens within a fraction of a second. More satellites are being put up all the time to meet the increasing demand for telecommunications.

The downlink from the satellite is easily intercepted. It is not a narrow beam, but a microwave signal that goes out in many directions. The higher the satellite, the larger the area on earth from which that radio wave can be received and, therefore, intercepted. For many satellites, the satellite "footprint," or area in which the satellite signals can be received on earth, is a couple thousand miles in diameter. The footprint can be reduced by lowering the satellite orbit or increasing the size of the satellite antenna, but the signals can still be received over a wide area.

Anyone within this footprint with a satellite dish and some readily available equipment can pick up the signal in the same way that a backyard satellite dish pulls in television signals. Interception of satellite signals can be done from embassies or other foreign-owned buildings, from ships at sea off the coast, or from a foreign base. Satellite communications to and from most areas of the United States are vulnerable to interception from one or more of these locations.

Land-based Microwave

Land-based microwave used to be a major means of transmitting long distance communications across the country. Now, long haul communications increasingly go via satellite or fiber optic land lines.

Land-based microwave is used mainly for traffic over short distances or between a local phone office and the nearest major satellite or land line link.

Land-based microwave transmissions are relayed from one tower to another. The towers are placed at about 25- to 30-mile intervals, because the signals go mainly in a straight line and don't follow the curvature of the earth. (You can often see the towers as you drive along an interstate highway.)

Like satellite communications, land-based microwave communications are easily intercepted by anyone within range using readily available equipment. One security weakness of all microwave transmissions, whether land-based or via satellite, is that the beams have "side lobes" or "spill" along the full distance between relay points. Using a well-aimed parabolic dish antenna, it is possible to intercept the signal from the side if there is direct line of sight to a section of the beam.

Many foreign embassies, consulates, trade offices, and foreign-owned office buildings and residences in the United States are located in areas that provide opportunities to intercept land-based microwave as well as satellite signals. Rooftop antennas of foreign offices in Washington DC, New York, San Francisco and elsewhere sometimes indicate which countries are actively monitoring U.S. communications.

Tapping Landlines

A tap on a phone line allows an eavesdropper to monitor or record all conversations on that line. Telephone taps come in many varieties. Contrary to some popular belief, a sophisticated phone tap is unlikely to be noticed by the phone user and may not be apparent even to a professional technical security countermeasures team using the latest equipment.

Consider the miles of telephone lines between your phone and the telephone company's central office. Conversations can be intercepted at any point along this path by several techniques. Sophisticated devices may be attached to or placed in or near communications equipment and cables. The tap may include a miniature transmitter that broadcasts the signal to a nearby listening post, a switch that allows monitoring from another line, or a voice-actuated recorder.

The limiting factor is that the installer of a telephone tap must somehow gain physical access to the telephone cables, terminals, or switching equipment for a brief period of time. In some cases the physical access may be readily available—for example, if the customer service box is located on the outside of a home or other target building. In other cases, a member of an unescorted cleaning crew might be recruited to provide access to the cables in a large office building, or a telephone repairman might be recruited to provide direct access to the lines or to a switching station.

In tapping phone lines, a local security service that can tap lines legally has a huge advantage over anyone who might try to do so without official support. American government and business offices overseas must assume their telephone lines are tapped, as this is a common practice. The capability is certainly there to tap any telephone, fax, e-mail, computer, or other form of electronic communication that might carry information of potential interest. Large volumes of tapes can be screened by computer programs that search for key words. Artificial intelligence algorithms can pick out the conversations most likely to contain useful information.

Fiber-optic cables are gradually replacing copper wire as a transmission media for both inside and outside wiring. While not as vulnerable as copper cable to simple methods of attack, fiber-optic cables are nonetheless vulnerable. Devices are readily available to extract information from cable previously billed by some as tap proof.

References:

1. "Reporter Pleads Guilty to Felony Charges in Chiquita Matter," *The Wall Street Journal*, Sept. 25, 1998, p. B4.

2. FBI Director Louis Freeh, Statement before the Senate Select Committee on Intelligence, January 28, 1998. Also PR Newswire, Former Sales Manager Charged in Voice Mail Scam, November 5, 1996.

3. Paul F. Barry & Charles L. Wilkinson (Trident Data Systems), "Invasion of Privacy and 90s Technologies," *Security Awareness Bulletin*, No. 2-96. Richmond, VA: Department of Defense Security Institute, August 1996.

4. Richard J. Heffernan, "Who's on the Line?" *Security Awareness Bulletin* 2-96, Department of Defense Security Institute, August 1996. Reprinted from *Security Management*, Vol. 36, September 1992.

5. Much of this description of the mechanics of intercepting microwave and satellite communications is from an article by

Senator Daniel Patrick Moynihan, "Privacy Disappears as America is Plagued by 'Bugs:' To the Soviets, All of America is a Party Line, as their Devices Tap Phone Communications." Published in *Popular Mechanics* and reprinted in *Orange County Register*, April 14, 1987.

Chapter 45

Wireless Intercom, Baby Monitor, Cordless Phone, and Cellular Surveillance Vulnerabilities

Wireless Intercoms and Baby Monitors[1]

Home wireless intercoms are transmitter and receiver sets that use home AC wiring for two purposes. First, they derive power for the units from the AC current. Second, they use the same AC wiring as a form of antenna to transmit and receive the signals from one unit to another. This is called "carrier current" technology.

Many wireless intercom owners may not know that their intercom broadcasts beyond the limits of their home. The signals can be picked up accidentally or deliberately by anyone nearby who has a wireless intercom operating on the same frequency or anyone with a frequency scanner looking for that type of emission. Moreover, the carrier current signals can be carried over great distances by wire. The power companies use the same carrier current technology to obtain information

---

This chapter contains text from "Wireless Intercoms & Baby Monitors," from an undated website produced by the Counterintelligence Training Academy, Nonproliferation and National Security Institute, Department of Energy (DOE), http://www.smdc.army.mil/Intelligence/Security%20Guide/V2comint/Intercom.htm, downloaded March 2002; "Cordless Phones," from an undated website produced by the Counterintelligence Training Academy, Nonproliferation and National Security Institute, Department of Energy (DOE), http://www.smdc.army.mil/Intelligence/Security%20Guide/V2comint/Cordless.htm, downloaded March 2002; and "Cellular Phones," from an undated website produced by the Counterintelligence Training Academy, Nonproliferation and National Security Institute, Department of Energy (DOE), http://www.smdc.army. mil/Intelligence/Security%20Guide/V2comint/Cellular.htm, downloaded March 2002.

that allows them to operate "load management" or "kilowatchers" systems to manage the power to large groups of homes.

Baby monitors also amount to self-installed bugging devices that transmit signals well outside the confines of the home where they are placed. Some are hooked into the home's AC wiring and use the same carrier current technology discussed above. Others transmit on common radio frequencies. The baby monitors are very effective in alerting parents to crying or other sounds in another part of their home. They provide peace of mind for new parents, but the potential privacy issues need to be recognized.

Many hotels and motels offer a version of the baby monitor to guests who wish to use the club or guest services, while still minding the toddlers asleep in their hotel room. This service is an integral part of the hospitality features incorporated into the hotel phone system. That means it is possible from a central location to activate the microphone within the telephone set within any room at any time. Consider the implications of this built-in surveillance system for government or business employees who think they can have confidential discussions in their hotel room.

Cordless Phones

Cordless telephones are great little step savers, but can be bad for privacy and security as the transmissions can be received up to a mile away. With the cheapest analog phones, anyone with a radio scanner in the general neighborhood of your phone can tune into your telephone's transmitting frequency and listen in on your conversations. Conversations may also be overheard on other phones or even picked up on baby monitors.

Cordless microphones used at meetings and conferences present a similar problem. They transmit crystal clear audio to any outsider who may have set up a listening post within a range of about a quarter mile. Many speakers prefer the mobility and flexibility provided by a cordless microphone, but such a microphone should not be used for any presentation of sensitive information.

The more you pay for a cordless phone, the more you get in security as well as range and sound quality. Digital phones are more secure because a casual or accidental eavesdropper would hear the digital noise rather than voice (much like the shrill sound you hear when you dial a phone number and get a fax machine rather than a person). Any radio hobbyist can, however, readily obtain equipment to convert the digital transmission back to voice.

Most secure are the digital spread spectrum 900 MHz and 2.4 GHz cordless phones. These use a broad bandwidth that cannot be monitored effectively by a typical radio frequency scanner. As with any radio transmission, however, it can be monitored by a determined professional.

Additionally, a cordless phone involves some financial risk. Anyone with a "like" cordless telephone can attempt to steal your dial tone and place a call leaving you with the bill. It works like this: a "bad guy" grabs his own cordless telephone and takes a drive in his car. He drives up and down the residential streets in your neighborhood with his telephone "on." When his telephone receives a dial tone, he has effectively located a telephone base unit that is compatible with his telephone. He then can dial any number he wants and can talk for virtually hours leaving the bill to his unsuspecting host. Even if his host were to pick up the telephone and discover someone else on the line, the host may not know how their telephone is being attacked (or know to look outside for someone parked on the street).[1]

Many cordless telephones are designed with security code options. However, the factory-set codes are well known, and most people don't bother to change them from the original factory setting. In the scenario above, the security code doesn't really matter. The bad guy merely drives until his telephone and code matches with some host with the same factory-set code. This still results in invasion of privacy and possibly an unexpected telephone bill.

Cellular Phones

Your cellular telephone has three major security vulnerabilities:

- Vulnerability to monitoring of your conversations while using the phone.
- Vulnerability of your phone being turned into a microphone to monitor conversations in the vicinity of your phone while the phone is inactive.
- Vulnerability to "cloning," or the use of your phone number by others to make calls that are charged to your account.

Before discussing these vulnerabilities, here is a brief tutorial on how cellular phones function. They send radio frequency transmissions through the air on two distinct channels, one for voice communications and the other for control signals. When a cellular telephone is first turned on, it emits a control signal that identifies itself to a cell site by broadcasting its mobile identification number (MIN) and electronic serial number (ESN), commonly known as the "pair."

When the cell site receives the pair signal, it determines if the requester is a legitimate registered user by comparing the requestor's pair to a cellular subscriber list. Once the cellular telephone's pair has been recognized, the cell site emits a control signal to permit the subscriber to place calls at will. This process, known as anonymous registration, is carried out each time the telephone is turned on or picked up by a new cell site.

Vulnerability to Monitoring

All cellular telephones are basically radio transceivers. Your voice is transmitted through the air on radio waves. Radio waves are not directional—they disperse in all directions so that anyone with the right kind of radio receiver can listen in.

Although the law provides penalties for the interception of cellular telephone calls, it is easily accomplished and impossible to detect. Radio hobbyists have web sites where they exchange cell phone numbers of "interesting" targets. Opportunistic hobbyists sometimes sell their best "finds." Criminal syndicates in several major U.S. metropolitan areas maintain extensive cell phone monitoring operations.

Cell phones operate on radio frequencies that can be monitored by commonly available radio frequency scanners.

- If the cellular system uses analog technology, one can program a phone number, or a watch list of phone numbers, into a cell-monitoring device that automatically turns on a voice-activated tape recorder whenever one of the watch listed numbers is in use. Computer assisted, automatic monitoring allows monitoring a specific phone 24 hours a day, as the target moves from cell to cell, without any human assistance.
- If the cellular system uses newer digital technology, it is possible for a price affordable by most radio hobbyists to buy a digital data interpreter that connects between a scanner radio and a personal computer. The digital data interpreter reads all the digital data transmitted between the cellular site and the cellular phone and feeds this information into the computer.[3]

It is easy for an eavesdropper to determine a target's cellular phone number, because transmissions are going back and forth to the cellular site whenever the cell phone has battery power and is able to receive a call. For a car phone, this generally happens as soon as the ignition is turned on. Therefore, the eavesdropper simply waits for the target to leave his or her home or office and start the car. The initial transmission to the cellular site to register the active system is picked up immediately by the scanner, and the number can be entered automatically into a file of numbers for continuous monitoring.

One of the most highly publicized cases of cellular phone monitoring concerned former Speaker of the House of Representatives Newt Gingrich. A conference call between Gingrich and other Republican leaders was "accidentally" overheard and then taped. The conversation concerned Republican strategy for responding to Speaker Gingrich's pending admission of ethics violations being investigated by the House Ethics Committee. The intercepted conversation was reported in the New York Times and other newspapers.[2]

Pagers have similar vulnerabilities. In 1997, police arrested officials of a small New Jersey company, Breaking News Network, that was monitoring pager messages to New York City leaders and police, fire, and court officials, including messages considered too sensitive to send over the police radio. They were selling the information to newspaper and television reporters. The offenses carry a penalty of up to five years in prison and fines of $250,000 for each offense.[4]

Vulnerability to Being Used as a Microphone

A cellular telephone can be turned into a microphone and transmitter for the purpose of listening to conversations in the vicinity of the phone. This is done by transmitting to the cell phone a maintenance command on the control channel. This command places the cellular telephone in the "diagnostic mode." When this is done, conversations in the immediate area of the telephone can be monitored over the voice channel.[5]

The user doesn't know the telephone is in the diagnostic mode and transmitting all nearby sounds until he or she tries to place a call. Then, before the cellular telephone can be used to place calls, the unit has to be cycled off and then back on again. This threat is the reason why cellular telephones are often prohibited in areas where classified or sensitive discussions are held.

Vulnerability to Cloning

Cellular telephone thieves don't steal cellular telephones in the usual sense of breaking into a car and taking the telephone hardware.

Instead, they monitor the radio frequency spectrum and steal the cell phone pair as it is being anonymously registered with a cell site.

Cloning is the process whereby a thief intercepts the electronic serial number (ESN) and mobile identification number (MIN) and programs those numbers into another telephone to make it identical to yours. Once cloned, the thief can place calls on the reprogrammed telephone as though he were the legitimate subscriber.

Cloning resulted in approximately $650 million dollars worth of fraudulent phone calls in 1996. Police made 800 arrests that year for this offense.[6] Each day more unsuspecting people are being victimized by cellular telephone thieves. In one case, more than 1,500 telephone calls were placed in a single day by cellular phone thieves using the number of a single unsuspecting owner.[1]

The ESN and MIN can be obtained easily by an ESN reader, which is like a cellular telephone receiver designed to monitor the control channel. The ESN reader captures the pair as it is being broadcast from a cellular telephone to a cell site and stores the information into its memory. What makes this possible is the fact that each time your cellular telephone is turned on or used, it transmits the pair to the local cellular site and establishes a talk channel. It also transmits the pair when it is relocated from one cell site to another.

Cloning occurs most frequently in areas of high cell phone usage—valet parking lots, airports, shopping malls, concert halls, sports stadiums, and high-congestion traffic areas in metropolitan cities. No one is immune to cloning, but you can take steps to reduce the likelihood of being the next victim.

Cellular Phone Security Measures

The best defense against these three major vulnerabilities of cell phones is very simple—do not use the cell phone. If you must use a cell phone, you can reduce the risk by following these guidelines:

- Because a cellular phone can be turned into a microphone without your knowledge, do not carry a cellular phone into any classified area or other area where sensitive discussions are held. (This is prohibited in many offices that handle classified or sensitive information.)
- Turn your cellular telephone on only when you need to place a call. Turn it off after placing the call. Do not give your cellular phone number to anyone and don't use your cell phone for receiving calls, as that requires leaving it on all the time. Ask your

friends and associates to page you if they need to talk with you. You can then return the page by using your cellular telephone.

- Do not discuss sensitive information on a cellular phone. When you call someone from your cell phone, consider advising them you are calling from a cell phone that is vulnerable to monitoring, and that you will be speaking generally and not get into sensitive matters.
- Do not leave your cellular telephone unattended. If your cell phone is vehicle-mounted, turn it off before permitting valet parking attendants to park the car, even if the telephone automatically locks when the car's ignition is turned off.
- Avoid using your cellular telephone within several miles of the airport, stadium, mall, or other heavy traffic locations. These are areas where radio hobbyists use scanners for random monitoring. If they come across an interesting conversation, your number may be marked for regular selective monitoring.
- If your cellular service company offers personal identification numbers (PIN), consider using one. Although cellular PIN services are cumbersome and require that you input your PIN for every call, they are an effective means of thwarting cloning.

References

1. Paul F. Barry & Charles L. Wilkinson (Trident Data Systems), "Invasion of Privacy and 90s Technologies," *Security Awareness Bulletin*, No. 2-96. Richmond, VA: Department of Defense Security Institute, August 1996.
2. Jessica Lee, "Focus Shifts from Gingrich to Taped Call," *USA Today*, Jan. 14, 1997, p. 5A.
3. "How O.J. Simpson was Tracked in his Bronco by Los Angeles Law Enforcement," *U.S. Scanner News*, February 1995.
4. Stephanie Mehta, "Prosecutors Charge Company for Spying on Pager Messages," *The Wall Street Journal*, August 28, 1997, p. A6.
5. "Just How Secure Is Your Cellular Phone?" article in *National Reconnaissance Organization Newsletter*,1997.
6. "Running Cell-Phone Pirates Aground," *Business Week*, October 27, 1997, p. 8.

Chapter 46

A Guide to Federal Law on Electronic Surveillance

1. What Is a Wiretap, a Pen Register, and a Trap-and-trace Device?

Wiretap: A wiretap is a device that acquires the content of an oral, wire or electronic communication (but not including telephone switchboards or hearing aids). 18 U.S.C. §2510(4). The "content" of a communication means any information concerning the substance, purport or meaning of the communication. 18 United States Code (U.S.C.) §2510(8).

Pen Register: A pen register is a device attached to a telephone line which records all telephone numbers dialed from a telephone on that line. 18 U.S.C. §3127(3).

Tap-and-Trace Device: A trap-and-trace device is a device attached to a telephone line which records the number of each telephone dialing into that telephone line. 18 U.S.C. §3127(4). For purposes of the federal law of electronic surveillance, pen registers and tap-and-trace devices are treated the same.

"Wiretap FAQ: a Guide to Federal Law on Electronic Surveillance," Privacy Foundation, http://www.privacyfoundation.org/resources/wiretap.asp.

2. What Is the Difference between an Oral Communication, a Wire Communication, and an Electronic Communication?

Oral communication: A human utterance made when the speaker reasonably expects that the utterance is not subject to interception. 18 U.S.C. §2510(2).

Wire communication: A human utterance transmitted in whole or in part over wire furnished by a telecommunications company. 18 U.S.C. §2510(1).

Electronic communication: A transmission of information not containing the human voice, made, in whole or in part, by wire, radio, electromagnetic, photoelectronic, or photo-optical system (but not including tone-only pagers, tracking devices, or electronic funds transfer information stored by a financial institution). 18 U.S.C. §2510(12).

3. Which Federal Government Officials Authorize an Application for an Interception Order?

Wiretap: For wire or oral communications, only specified, high-level Justice Department attorneys. 18 U.S.C. §2516(1). For electronic communications, any federal government attorney. 18 U.S.C. §2516(3).

Pen Register: Any federal government attorney. 18 U.S.C. §3122(a)(1).

4. For What Types of Crimes May an Interception Order Be Requested?

Wiretap: For wire or oral communications, only serious felonies specifically identified by statute. These crimes typically involve threats to national security, serious bodily harm or death, organized crime, or conspiratorial conduct. 18 U.S.C. §2516(a)-(p). For electronic communication, any federal felony. 18 U.S.C. §2516(3).

Pen Register: Any crime. 18 U.S.C.§3122(2).

5. What Information must the Government Include in an Application for an Interception Order?

Wiretap: For all types of communications, the application must be made under oath and state the following: (a) the identity of the applicant and the person authorizing the application, (b) the offense that has been, is being, or is about to be committed, (c) the location of the wiretap, (d) the type of communication to be intercepted, (e) the target of the wiretap, (f) an explanation why other investigative procedures have not succeeded or are unlikely to succeed, (g) the number of days the wiretap will last, (h) a description of any previous wiretaps of the same target, and (i) any additional information required by the judge to whom the application is made. 18 U.S.C. §2518(1), (2).

Pen Register: The application must be made under oath and state the following: (a) the identity of the applicant and the law enforcement agency conducting the investigation, and (b) that the information likely to be obtained is relevant to an ongoing criminal investigation being conducted by the agency. 18 U.S.C. §3122(b).

6. Who Can Grant an Application by the Federal Government for an Interception Order?

Wiretap: A federal judge. 18 U.S.C. §2516(1).

Pen Register: A federal judge or a federal magistrate judge. 18 U.S.C. §3123(a).

7. What Standard Must the Government Satisfy before the Application Will Be Granted?

Wiretap: The judge must determine that (a) there is probable cause that the target has committed, is committing, or is about to commit, a crime, and (b) other investigative procedures have failed or will not succeed. 18 U.S.C. §2518(3).

Pen Register: The judge or magistrate judge must determine only that the government has provided the information required to be included in the application. 18 U.S.C. §3123(a).

8. What Must Be Stated in the Order Authorizing the Interception?

Wiretap: The order must specify the following: (a) the identity of the target, (b) the location of the wiretap, (c) the type of communications to be intercepted and the particular offense to which the communications relates, (d) the identity of the agency authorized to intercept the communications and of the attorney authorizing the application, (e) the period of time during which interception is permitted and whether the interception must terminate when the communications sought are first obtained. 18 U.S.C. §2518(4).

The order also must state that the interception (a) shall be executed as soon as practicable, (b) shall be conducted in such a way as to minimize the interception of communications not within the scope of the order, and (c) must terminate upon attainment of the objective of the interception or in thirty days, whichever is sooner. 18 U.S.C. §2518(5).

Pen Register: The order must specify the following: (a) the person in whose name the telephone line to which the device will be attached is listed, (b) the identity of the target of the investigation, (c) the telephone number and physical location of the telephone line to which the device will be attached, and (d) a statement of the offense to which the telephone numbers likely to be obtained relate. 18 U.S.C. §3123(b).

9. For What Period of Time May the Interception Be Authorized?

Wiretap: The judge may authorize interceptions for up to 30 days. The government may request an extension of up to 30 days in accordance with the procedures for the initial application. In all events, the interception must end when the objective is attained. 18 U.S.C. §2518(5).

Pen Register: The judge or magistrate judge may authorize the installation and use of the device for up to 60 days. The government may request an extension of up to 60 days in accordance with the procedures for the initial application. 18 U.S.C. §3123(c).

10. Under What Circumstances May the Federal Government Intercept without First Obtaining Court Authorization?

Wiretap: A law enforcement officer specially designated by the Attorney General, the Deputy Attorney General or the Associate Attorney General may authorize an interception without prior court authorization if that person reasonably determines that (a) an emergency situation exists involving immediate danger of death or serious physical injury to any person, a threat to national security, or organized crime, (b) an interception is necessary before court authorization can be obtained, and (c) there are grounds for obtaining court authorization. An application for judicial authorization must be made within 48 hours of the commencement of the interception. 18 U.S.C. §2518(7).

The interception shall terminate when the communication sought is obtained or when the application for a wiretap order is denied, whichever is sooner. If no wiretap order is granted, the interception shall be treated as though it had been obtained unlawfully (i.e., it is subject to suppression) and the target shall be provided with notice of the interception and an inventory of the communications intercepted. 18 U.S.C. §2518(7).

Pen Register: Any law enforcement officer specially designated by the Attorney General, the Deputy Attorney General, the Associate Attorney General, any Assistant Attorney General, any acting Assistant Attorney, any Deputy Assistant Attorney General, or any acting Deputy Assistant Attorney General may authorize the use and installation of a pen register if that person reasonably determines that (a) an emergency exists involving immediate danger of death or serious physical injury to any person or involving organized crime, (b) installation and use of a pen register is required before court authorization can be obtained, and (c) there are grounds for obtaining court authorization. An application for judicial authorization must be made within 48 hours after the installation of the pen register. 18 U.S.C. §3125(a).

The use of the pen register shall cease when the information sought is obtained, when judicial authorization is denied, or within 48 hours after installation of the device, whichever is sooner. 18 U.S.C. §3125(b).

11. When Do the People Whose Communications Are Intercepted Receive Notice of the Interceptions from the Federal Government?

Wiretap: No later than 90 days after completion of the wiretap, notice must be given to the target of the investigation and, if the authorizing judge so determines, to other persons whose communications were intercepted in accordance with the wiretap order. The notice must include an inventory of all communications which were intercepted. 18 U.S.C. §2518(8)(d).

Pen Register: There is no provision for notification.

12. What Remedies Are Available to a Person Whose Communications Are Intercepted in Violation of Federal Law?

Wiretap: For oral and wire communications, but not electronic communications, the victim of an unlawful interception may move to suppress, in any hearing or trial, all unlawfully intercepted communications and any evidence derived therefrom. 18 U.S.C. §2515. For oral, wire and electronic communications, the victim may recover damages in a civil lawsuit. 18 U.S.C. §2520.

Pen Register: None.

13. What Is the Range of Criminal Penalties That Can Be Imposed on a Federal Government Official Who Intercepts in Violation of Federal Law?

Wiretap: A fine and imprisonment of up to five years. 18 U.S.C. §2511(4)(a).

Pen Register: A fine and imprisonment of up to one year. 18 U.S.C. §3121(d).

Chapter 47

Communications Assistance for Law Enforcement Act (CALEA)

Introduction

The purpose of this chapter is to provide a brief summary of the importance of lawfully-authorized electronic surveillance to the law enforcement community and describe some of the electronic surveillance techniques in use today. In addition, this chapter provides an historical perspective as to the need for CALEA and describes the efforts of the CALEA Implementation Section (CIS) to fulfill the legislative responsibilities of the Attorney General.

Importance of Electronic Surveillance

Electronic surveillance is one of the most valuable tools in law enforcement's crime fighting arsenal. In many instances, criminal activity has either been thwarted, or if crimes have been committed, the criminals have been apprehended as a result of lawfully authorized electronic surveillance. Electronic surveillance is strictly and carefully regulated by federal statute. Courts only authorize interceptions after an exhaustive demonstration of need by law enforcement.

The use of lawfully authorized electronic surveillance continues to increase in importance to law enforcement as telecommunications

---

Excerpted from "Communications Assistance for Law Enforcement Act (CALEA) Guide," Federal Bureau of Investigation (FBI), CALEA Implementation Section (CIS), http://www.askcalea.com/pdf/CALEA_Guide.pdf, June 2001.

systems become cornerstones of everyday life: dependence on telecommunications for business and personal use has increased dramatically; computers and data services have become increasingly important to consumers; and the nation has become enthralled with mobile communications.

Types of Electronic Surveillance

Three primary techniques of lawfully authorized electronic surveillance are available to law enforcement: pen registers, trap and trace devices, and content interceptions. The first two, pen registers and trap and trace devices, which account for the vast majority of lawfully authorized electronic surveillances, record/decode various types of dialing and signaling information utilized in processing and routing the communication, such as the signals that identify the numbers dialed (i.e., outgoing) or the originating (i.e., incoming) number of a telephone communication. The third, and more comprehensive form of lawfully authorized electronic surveillance, includes not only the acquisition of call-identifying, or dialed number information, but also the interception of communications content.

Although electronic surveillance is crucial to effective law enforcement, it is used sparingly. The federal government, District of Columbia, Guam, Puerto Rico, Virgin Islands, and 42 states allow its use, but only in the investigation of felony offenses, such as kidnaping, extortion, murder, illegal drug trafficking, organized crime, terrorism, and national security matters, and only when other investigative techniques either can not provide the needed information or would be too dangerous.

In addition to the 42 states that allow the three forms of electronic surveillance discussed above, three states (Maine, Michigan, and South Carolina) permit the use of pen registers and trap and trace devices for the surveillance of call-identifying information. The five remaining states (Alabama, Arkansas, Kentucky, Montana, and Vermont) do not permit any of the three forms of electronic surveillance discussed above. However, federal law enforcement agencies have the authority to conduct electronic surveillance throughout the entire country.

What Is CALEA?

> *The purpose of [CALEA] is to preserve the Government's ability, pursuant to court order or other lawful authorization, to intercept communications involving advanced technologies such as digital or wireless transmission modes, or features and services*

such as call forwarding, speed dialing and conference calling, while protecting the privacy of communications and without impeding the introduction of new technologies, features, and services.

To [e]nsure that law enforcement can continue to conduct authorized wiretaps in the future, the bill requires telecommunications carriers to ensure their systems have the capability to: (1) isolate expeditiously the content of targeted communications transmitted by the carrier within the carrier's service area; (2) isolate expeditiously information identifying the origin and destination of targeted communications; (3) provide intercepted communications and call identifying information to law enforcement so they can be transmitted over lines or facilities leased by law enforcement to a location away from the carrier's premises; and (4) carry out intercepts unobtrusively, so targets are not made aware of the interception, and in a manner that does not compromise the privacy and security of other communications.

[H. Rep. No. 103-827, 103d Cong., 2d Sess. 9 (1994).]

Why Was CALEA Passed?

Passage of the Communications Assistance for Law Enforcement Act was a logical and necessary development of our Nation's electronic surveillance laws—originating in the enactment of Title III of the Omnibus Crime Control and Safe Streets Act of 1968. Title III formed the foundation for communications privacy and law enforcement's electronic surveillance authority. In response to continued advances in telecommunications technology, Congress passed the Electronic Communications Privacy Act (ECPA) of 1986, which confirmed law enforcement's electronic surveillance authority to include emerging technologies and services such as electronic mail, cellular telephones, and paging devices.

Following enactment of the ECPA, advancements in telecommunications technology continued to challenge and, in some cases, thwart law enforcement's electronic surveillance capability. What was once a simple matter of attaching wires to terminal posts now requires expert assistance from telecommunications service providers. Examples of technological changes that impeded law enforcement's efforts to conduct lawfully authorized electronic surveillance include:

- New, switch-based features and services (e.g., call forwarding, call transfer, one number service, follow-me service),

- Conversion from analog to digital transmission modes between the subscriber and carrier,
- Cellular, Personal Communications Services (PCS), and other nonwireline technologies that do not allow law enforcement access through a local loop.

Although Title III required telecommunications carriers to provide "any assistance necessary to accomplish an electronic interception," the question of whether companies had an obligation to *design* their networks such that they did not impede a lawfully authorized interception had not been decided.

In October 1994, at the request of the nation's law enforcement community, Congress responded to this dilemma by enacting CALEA, which clarifies the scope of a carrier's duty in effecting court-approved electronic surveillance.

Implementation of CALEA

The implementation of CALEA requires the active participation and cooperation of a variety of parties. In addition to the law enforcement community, there are a significant number of stakeholders and influential players that need to work together to ensure that the introduction of sophisticated telecommunications technologies into the nation's networks does not erode law enforcement's ability to conduct lawfully authorized electronic surveillance. Telecommunications carriers; manufacturers of telecommunications equipment and providers of support services; industry associations; privacy advocates; the Federal Communications Commission and state utility commissions; the Courts, and the law enforcement community all play a critical role in the implementation of CALEA.

What Is CIS?

CIS is a Section within the FBI's Laboratory Division. The Laboratory Division is that part of the FBI that has, as part if its organizational mission, the obligation to ". . . provide effective collection, surveillance, and tactical communications systems to support investigative and intelligence priorities." The inclusion of CALEA implementation responsibilities into this part of the FBI allows CIS to draw upon the extensive telecommunications knowledge and electronic surveillance experience of the FBI.

Why Was CIS Created?

CIS was established in response to the delegation of CALEA implementation responsibilities to the FBI by the Attorney General.

Responsibilities and Authority

CIS spearheads CALEA implementation efforts by fulfilling the responsibilities assigned to the Attorney General through consultation with the telecommunications industry and privacy advocates. CIS represents the interests of the entire law enforcement community before Congress, the Federal Communications Commission, and other government agencies involved in the implementation of CALEA, and the telecommunications industry.

CIS views the "implementation of CALEA" to mean all explicit and implicit mandates and responsibilities contained within the legislation. These responsibilities apply to *all* telecommunications technologies and services that are mandated to comply with the requirements of CALEA.

To date, CIS has made significant implementation progress with traditional wireline local exchange, cellular, and broadband PCS services as evidenced by the availability of technical solutions for a majority of equipment currently in use by carriers providing those services.

Other telecommunications technologies and services that CIS is currently pursuing implementation include Specialized Mobile Radio and Enhanced Specialized Mobile Radio; traditional, two-way paging, and ancillary paging services; mobile satellite services; and Internet protocol-based telecommunications.

Chapter 48

Frequently Asked Questions about Echelon

Q—What Is Project ECHELON?

ECHELON is the term popularly used for an automated global interception and relay system operated by the intelligence agencies in five nations: the United States, the United Kingdom, Canada, Australia and New Zealand (it is believed that ECHELON is the code name for the portion of the system that intercepts satellite-based communications). While the United States National Security Agency (NSA) takes the lead, ECHELON works in conjunction with other intelligence agencies, including the Australian Defence Signals Directorate (DSD). It is believed that ECHELON also works with Britain's Government Communications Headquarters (GCHQ) and the agencies of other allies of the United States, pursuant to various treaties.[1]

These countries coordinate their activities pursuant to the UKUSA agreement, which dates back to 1947. The original ECHELON dates back to 1971. However, its capabilities and priorities have expanded greatly since its formation. According to reports, it is capable of intercepting and processing many types of transmissions, throughout the globe. In fact, it has been suggested that ECHELON may intercept as many as 3 billion communications everyday, including phone calls, e-mail messages, Internet downloads, satellite transmissions, and so on.[2] The ECHELON system gathers all of these transmissions

---

indiscriminately, then distills the information that is most heavily desired through artificial intelligence programs. Some sources have claimed that ECHELON sifts through an estimated 90 percent of all traffic that flows through the Internet.[3]

However, the exact capabilities and goals of ECHELON remain unclear. For example, it is unknown whether ECHELON actually targets domestic communications. Also, it is apparently very difficult for ECHELON to intercept certain types of transmissions, particularly fiber communications.

Q—How Does ECHELON Work?

ECHELON apparently collects data in several ways. Reports suggest it has massive ground based radio antennae to intercept satellite transmissions. In addition, some sites reputedly are tasked with tapping surface traffic. These antennae reportedly are in the United States, Italy, England, Turkey, New Zealand, Canada, Australia, and several other places.[4]

Similarly, it is believed that ECHELON uses numerous satellites to catch "spillover" data from transmissions between cities. These satellites then beam the information down to processing centers on the ground. The main centers are in the United States (near Denver), England (Menwith Hill), Australia, and Germany.[5]

According to various sources, ECHELON also routinely intercepts Internet transmissions. The organization allegedly has installed numerous "sniffer" devices. These "sniffers" collect information from data packets as they traverse the Internet via several key junctions. It also uses search software to scan for web sites that may be of interest.[6]

Furthermore, it is believed that ECHELON has even used special underwater devices which tap into cables that carry phone calls across the seas. According to published reports, American divers were able to install surveillance devices on to the underwater cables. One of these taps was discovered in 1982, but other devices apparently continued to function undetected.[7]

It is not known at this point whether ECHELON has been able to tap fiber optic phone cables.

Finally, if the aforementioned methods fail to garner the desired information, there is another alternative. Apparently, the nations that are involved with ECHELON also train special agents to install a variety of special data collection devices. One of these devices is reputed to be an information processing kit that is the size of a suitcase. Another such item is a sophisticated radio receiver that is as small as a credit card.[8]

After capturing this raw data, ECHELON sifts through them using DICTIONARY. DICTIONARY is actually a special system of computers which finds pertinent information by searching for key words, addresses, etc. These search programs help pare down the voluminous quantity of transmissions which pass through the ECHELON network every day. These programs also seem to enable users to focus on any specific subject upon which information is desired.[9]

Q—If ECHELON Is So Powerful, Why Haven't I Heard about It Before?

The United States government has gone to extreme lengths to keep ECHELON a secret. To this day, the U.S. government refuses to admit that ECHELON even exists. We know it exists because both the governments of Australia (through its Defence Signals Directorate) and New Zealand have admitted to this fact.[10] However, even with this revelation, US officials have refused to comment.

This "wall of silence" is beginning to erode. The first report on ECHELON was published in 1988.[11] In addition, besides the revelations from Australia, the Scientific and Technical Options Assessment program office (STOA) of the European Parliament commissioned two reports which describe ECHELON's activities. These reports unearthed a startling amount of evidence, which suggests that Echelon's powers may have been underestimated. The first report, entitled "An Appraisal of Technologies of Political Control," suggested that ECHELON primarily targeted civilians.

This report found that:

> "The ECHELON system forms part of the UKUSA system but unlike many of the electronic spy systems developed during the cold war, ECHELON is designed for primarily non-military targets: governments, organizations and businesses in virtually every country. The ECHELON system works by indiscriminately intercepting very large quantities of communications and then siphoning out what is valuable using artificial intelligence aids like Memex to find key words. Five nations share the results with the US as the senior partner under the UKUSA agreement of 1948, Britain, Canada, New Zealand and Australia are very much acting as subordinate information servicers.
>
> "Each of the five centres supply 'dictionaries' to the other four of keywords, phrases, people and places to 'tag' and the tagged intercept is forwarded straight to the requesting country. Whilst

> there is much information gathered about potential terrorists, there is a lot of economic intelligence, notably intensive monitoring of all the countries participating in the GATT negotiations. But Hager found that by far the main priorities of this system continued to be military and political intelligence applicable to their wider interests. Hager quotes from a 'highly placed intelligence operatives' who spoke to the Observer in London. 'We feel we can no longer remain silent regarding that which we regard to be gross malpractice and negligence within the establishment in which we operate.' They gave as examples. GCHQ interception of three charities, including Amnesty International and Christian Aid. 'At any time GCHQ is able to home in on their communications for a routine target request,' the GCHQ source said. In the case of phone taps the procedure is known as Mantis. With telexes its called Mayfly. By keying in a code relating to third world aid, the source was able to demonstrate telex 'fixes' on the three organizations. With no system of accountability, it is difficult to discover what criteria determine who is not a target."[12]

A more recent report, known as Interception Capabilities 2000, describes ECHELON capabilities in even more elaborate detail.[13] The release of the report sparked accusations from the French government that the United States was using ECHELON to give American companies an advantage over rival firms.[14] In response, R. James Woolsey, the former head of the US Central Intelligence Agency (CIA), charged that the French government was using bribes to get lucrative deals around the world, and that US surveillance networks were used simply to level the playing field.[15] However, experts have pointed out that Woolsey missed several key points. For example, Woolsey neglected to mention alleged instances of economic espionage (cited in Intelligence Capabilities 2000) that did not involve bribery. Furthermore, many observers expressed alarm with Woolsey's apparent assertion that isolated incidents of bribery could justify the wholesale interception of the world's communications.[16]

The European Parliament formed a temporary Committee of Enquiry to investigate ECHELON abuses.[17] In May 2001, members of this committee visited the United States in an attempt to discover more details about ECHELON. However, officials from both the NSA and the US Central Intelligence Agency (CIA) canceled meetings that they had previously scheduled with the European panel. The committee's chairman, Carlos Coelho, said that his group was "very disappointed" with the apparent rebuffs; in protest, the Parliamentary representatives returned home a day early.[18]

Afterwards, the committee published a report stating that ECHELON does indeed exist and that individuals should strongly consider encrypting their e-mails and other Internet messages.[19] However, the panel was unable to confirm suspicions that ECHELON is used to conduct industrial espionage, due to a lack of evidence.[20] Ironically, the report also mentioned the idea that European government agents should be allowed greater powers to decrypt electronic communications, which was criticized by some observers (including several members of the committee) as giving further support to Europe's own ECHELON-type system.[21] The European Parliament approved the report, but despite the apparent need for further investigation, the committee was disbanded.[22] Nevertheless, the European Commission plans to draft a "roadmap" for data protection that will address many of the concerns aired by the EP panel.[23]

Meanwhile, after years of denying the existence of ECHELON, the Dutch government issued a letter that stated: "Although the Dutch government does not have official confirmation of the existence of Echelon by the governments related to this system, it thinks it is plausible this network exists. The government believes not only the governments associated with Echelon are able to intercept communication systems, but that it is an activity of the investigative authorities and intelligence services of many countries with governments of different political signature."[24] These revelations worried Dutch legislators, who had convened a special hearing on the subject. During the hearing, several experts argued that there must be tougher oversight of government surveillance activities. There was also considerable criticism of Dutch government efforts to protect individual privacy, particularly the fact that no information had been made available relating to Dutch intelligence service's investigation of possible ECHELON abuses.[25]

In addition, an Italian government official has begun to investigate Echelon's intelligence-gathering efforts, based on the belief that the organization may be spying on European citizens in violation of Italian or international law.[26]

Events in the United States have also indicated that the "wall of silence" might not last much longer. Exercising their Constitutionally created oversight authority, members of the House Select Committee on Intelligence started asking questions about the legal basis for NSA's ECHELON activities. In particular, the Committee wanted to know if the communications of Americans were being intercepted and under what authority, since US law severely limits the ability of the intelligence agencies to engage in domestic surveillance. When asked about its legal authority, NSA invoked the attorney-client privilege and refused to disclose the legal standards by which ECHELON might have conducted its activities.[27]

President Clinton then signed into law a funding bill which required the NSA to report on the legal basis for ECHELON and similar activities.[28] However, the subsequent report (entitled Legal Standards for the Intelligence Community in Conducting Electronic Surveillance) gave few details about Echelon's operations and legality.[29]

However, during these proceedings, Rep. Bob Barr (R-GA), who has taken the lead in Congressional efforts to ferret out the truth about ECHELON, stated that he had arranged for the House Government Reform and Oversight Committee to hold its own oversight hearings.[30]

Finally, the Electronic Privacy Information Center has sued the US Government, hoping to obtain documents which would describe the legal standards by which ECHELON operates.[31]

Q—What Is Being Done with the Information That ECHELON Collects?

The original purpose of ECHELON was to protect national security. That purpose continues today. For example, we know that ECHELON is gathering information on North Korea. Sources from Australia's DSD have disclosed this much because Australian officials help operate the facilities there which scan through transmissions, looking for pertinent material.[32] Similarly, the Spanish government has apparently signed a deal with the United States to receive information collected using ECHELON. The consummation of this agreement was confirmed by Spanish Foreign Minister Josep Pique, who tried to justify this arrangement on security grounds.[33]

However, national security is not Echelon's only concern. Reports have indicated that industrial espionage has become a part of Echelon's activities. While present information seems to suggest that only high-ranking government officials have direct control over Echelon's tasks, the information that is gained may be passed along at the discretion of these very same officials. As a result, much of this information has been given to American companies, in apparent attempts to give these companies an edge over their less knowledgeable counterparts.[34]

In addition, there are concerns that Echelon's actions may be used to stifle political dissent. Many of these concerns were voiced in a report commissioned by the European Parliament. What is more, there are no known safeguards to prevent such abuses of power.[35]

Q—Is There Any Evidence That ECHELON Is Doing Anything Improper or Illegal with the Spying Resources at Its Disposal?

ECHELON is a highly classified operation, which is conducted with little or no oversight by national parliaments or courts. Most of what is known comes from whistleblowers and classified documents. The simple truth is that there is no way to know precisely what ECHELON is being used for.

But there is evidence, much of which is circumstantial, that ECHELON (along with its British counterpart) has been engaged in significant invasions of privacy. These alleged violations include secret surveillance of political organizations, such as Amnesty International.[36] It has also been reported that ECHELON has engaged in industrial espionage on various private companies such as Airbus Industries and Panavia, then has passed along the information to their American competitors.[37] It is unclear just how far Echelon's activities have harmed private individuals.

However, the most sensational revelation was that Diana, Princess of Wales may have come under ECHELON surveillance before she died. As reported in the Washington Post, the NSA admitted that they possessed files on the Princess, partly composed of intercepted phone conversations. While one official from the NSA claimed that the Princess was never a direct target, this disclosure seems to indicates the intrusive, yet surreptitious manner by which ECHELON operates.[38]

What is even more disquieting is that, if these allegations are proven to be true, the NSA and its compatriot organizations may have circumvented countless laws in numerous countries. Many nations have laws in place to prevent such invasions of privacy. However, there are suspicions that ECHELON has engaged in subterfuge to avoid these legal restrictions. For example, it is rumored that nations would not use their own agents to spy on their own citizens, but assign the task to agents from other countries.[39] In addition, as mentioned earlier, it is unclear just what legal standards ECHELON follows, if any actually exist. Thus, it is difficult to say what could prevent ECHELON from abusing its remarkable capabilities.

Q—Is Everyone Else Doing What ECHELON Does?

Maybe not everyone else, but there are plenty of other countries that engage in the type of intelligence gathering that ECHELON performs. These countries apparently include Russia, France, Israel, India,

Pakistan and many others.[40] Indeed, the excesses of these ECHELON-like operations are rumored to be similar in form to their American equivalents, including digging up information for private companies to give them a commercial advantage.

However, it is also known that ECHELON system is the largest of its kind. What is more, its considerable powers are enhanced through the efforts of America's allies, including the United Kingdom, Canada, Australia, and New Zealand. Other countries don't have the resources to engage in the massive garnering of information that the United States is carrying out.

Notes

1. Development of Surveillance Technology and Risk of Abuse of Economic Information (An appraisal of technologies for political control), Part 4/4: The state of the art in Communications Intelligence (COMINT) of automated processing for intelligence purposes of intercepted broadband multi-language leased or common carrier systems, and its applicability to COMINT targeting and selection, including speech recognition, Ch. 1, para. 5, PE 168.184 / Part 4/4 (April 1999). See Duncan Campbell, Interception Capabilities 2000 (April 1999) (http://www.iptv reports.mcmail.com/stoa_cover.htm).
2. Kevin Poulsen, Echelon Revealed, ZDTV (June 9, 1999).
3. Greg Lindsay, The Government Is Reading Your E-Mail, *TIME DIGITAL DAILY* (June 24, 1999).
4. PE 168.184 / Part 4/4, supra note 1, Ch. 2, para. 32–34, 45–46.
5. Id. Ch. 2, para. 42.
6. Id. Ch. 2, para. 60.
7. Id. Ch. 2, para. 50.
8. Id. Ch. 2, para. 62–63.
9. An Appraisal of Technologies for Political Control, at 20, PE 166.499 (January 6, 1998). See Steve Wright, An Appraisal of Technologies for Political Control (January 6, 1998) (http://cryptome.org/stoa-atpc.htm).
10. Letter from Martin Brady, Director, Defence Signals Directorate, to Ross Coulhart, Reporter, Nine Network Australia 2

(Mar. 16, 1999) (on file with the author); see also Calls for inquiry into spy bases, *ONE NEWS* New Zealand (Dec. 28, 1999).

11. Duncan Campbell, Somebody's listening, *NEW STATESMAN*, 12 August 1988, Cover, pages 10–12. See Duncan Campbell, ECHELON: NSA's Global Electronic Interception, (last visited October 12, 1999) (http://jya.com/echelon-dc.htm).

12. PE 166.499, supra note 9, at 19–20.

13. PE 168.184 / Part 4/4, supra note 1.

14. David Ruppe, Snooping on Friends?, ABCNews.com (US) (Feb. 25, 2000) (http://abcnews.go.com/sections/world/dailynews/echelon000224.html).

15. R. James Woolsey, Why We Spy on Our Allies, *WALL ST. J.*, March 17, 2000. See also *CRYPTOME*, Ex-CIA Head: Why We Spy on Our Allies (last visited April 11, 2000) (http://cryptome.org/echelon-cia2.htm).

16. Letter from Duncan Campbell to the Wall Street Journal (March 20, 2000) (on file with the author). See also Kevin Poulsen, Echelon Reporter answers Ex-CIA Chief, SecurityFocus.com (March 23, 2000) (http://www.securityfocus.com/news/6).

17. Duncan Campbell, Flaw in Human Rights Uncovered, *HEISE TELEPOLIS*, April 8, 2000. See also *HEISE ONLINE*, Flaw in Human Rights Uncovered (April 8, 2000) (http://www.heise.de/tp/english/inhalt/co/6724/1.html).

18. Angus Roxburgh, EU investigators 'snubbed' in US, *BBC News*, May 11, 2001 (http://news.bbc.co.uk/hi/english/world/europe/newsid_1325000/1325186.stm).

19. Report on the existence of a global system for intercepting private and commercial communications (ECHELON interception system), PE 305.391 (July 11, 2001) (available in PDF or Word format at http://www2.europarl.eu.int).

20. Id.; see also E-mail users warned over spy network, *BBC News*, May 29, 2001 (http://news.bbc.co.uk/hi/english/world/europe/newsid_1357000/1357264.stm).

21. Steve Kettman, Echelon Furor Ends in a Whimper, *Wired News*, July 3, 2001 (http://www.wired.com/news/print/0,1294,44984,00.html).

22. European Parliament resolution on the existence of a global system for the interception of private and commercial communications (ECHELON interception system) (2001/2098(INI)), A5-0264/2001, PE 305.391/DEF (Sept. 5, 2001) (available at http://www3.europarl.eu.int); Christiane Schulzki-Haddouti, Europa-Parlament verabsciedet Echelon-Bericht, Heise Telepolis, Sept. 5, 2001 (available at http://www.heise.de/tp/); Steve Kettman, Echelon Panel Calls It a Day, *Wired News*, June 21, 2001 (http://www.wired.com/news/print/0,1294,44721,00.html).

23. European Commission member Erkki Liikanen, Speech regarding European Parliament motion for a resolution on the Echelon interception system (Sept. 5, 2001) (transcript available at http://europa.eu.int).

24. Jelle van Buuren, Dutch Government Says Echelon Exists, *Heise Telepolis*, Jan. 20, 2001 (available at http://www.heise.de/tp/).

25. Jelle van Buuren, Hearing On Echelon In Dutch Parliament, *Heise Telepolis*, Jan. 23, 2001 (available at http://www.heise.de/tp/).

26. Nicholas Rufford, Spy Station F83, *SUNDAY TIMES* (London), May 31, 1998. See Nicholas Rufford, Spy Station F83 (May 31, 1998) (http://www.sunday-times.co.uk/news/pages/sti/98/05/31/stifocnws01003.html?999).

27. H. Rep. No. 106-130 (1999). See *Intelligence Authorization Act for Fiscal Year 2000*, Additional Views of Chairman Porter J. Goss (http://www.echelonwatch.org/goss.htm).

28. *Intelligence Authorization Act for Fiscal Year 2000*, Pub. L. 106-120, Section 309, 113 Stat. 1605, 1613 (1999). See H.R. 1555 *Intelligence Authorization Act for Fiscal Year 2000* (Enrolled Bill (Sent to President)) http://www.echelonwatch.org/hr1555c.htm).

29. UNITED STATES NATIONAL SECURITY AGENCY, LEGAL STANDARDS FOR THE INTELLIGENCE COMMUNITY IN CONDUCTING ELECTRONIC SURVEILLANCE (2000) (http://www.fas.org/irp/nsa/standards.html).

30. House Committee to Hold Privacy Hearings, (August 16, 1999) (http://www.house.gov/barr/p_081699.html).

31. ELECTRONIC PRIVACY INFORMATION CENTER, PRESS RELEASE: LAWSUIT SEEKS MEMOS ON SURVEILLANCE OF AMERICANS; EPIC LAUNCHES STUDY OF NSA INTERCEPTION ACTIVITIES (1999). See also Electronic Privacy Information Center, EPIC Sues for NSA Surveillance Memos (last visited December 17, 1999) (http://www.epic.org/open_gov/foia/nsa_suit_12_99.html).

32. Ross Coulhart, Echelon System: FAQs and website links, (May 23, 1999).

33. Isambard Wilkinson, US wins Spain's favour with offer to share spy network material, *Sydney Morning Herald*, June 18, 2001 (http://www.smh.com.au/news/0106/18/text/world11.html).

34. PE 168.184 / Part 4/4, supra note 1, Ch. 5, para. 101–103.

35. PE 166.499, supra note 9, at 20.

36. Id.

37. PE 168.184 / Part 4/4, supra note 1, Ch. 5, para. 101-102; Brian Dooks, EU vice-president to claim US site spies on European business, *YORKSHIRE POST*, Jan. 30, 2002 (available at http://yorkshirepost.co.uk).

38. Vernon Loeb, NSA Admits to Spying on Princess Diana, *WASHINGTON POST*, December 12, 1998, at A13. See Vernon Loeb, NSA Admits to Spying on Princess Diana, *WASHINGTON POST*, A13 (December 12, 1998) (http://www.washington post.com/wp-srv/national/daily/dec98/diana12.htm).

39. Ross Coulhart, Big Brother is listening, (May 23, 1999).

40. PE 168.184 / Part 4/4, supra note 1, Ch. 1, para. 7.

Part Six

Additional Help and Information

Chapter 49

Glossary of Telecommunications Security Terms

A

access charge: A fee charged subscribers or other telephone companies by a local exchange carrier for the use of its local exchange networks.

advance-fee loan scheme: In advance-fee loan schemes, persons with bad credit are promised a loan in return for a fee paid in advance.

analog signal: A signaling method that uses continuous changes in the amplitude or frequency of a radio transmission to convey information.

"Glossary of Telecommunications Terms," Federal Communications Commission (FCC), http://www.fcc.gov/glossary.html, July 31, 2002. Terms from this glossary were also compiled from the following government sources: Counterintelligence Training Academy; Department of Justice (DOJ); Federal Communications Commission (FCC); Department of Navy Information Technology (IT) Umbrella Program. Terms from this glossary were also compiled from the following copyrighted sources: [1] Copyright 2002, American Civil Liberties Union. Reprinted with permission of the American Civil Liberties Union, http://www.aclu.org; [2] Reprinted with permission from the Electronic Frontier Foundation, http://www.eff.org; [3] ©2002 How Stuff Works, Inc. Reprinted with permission, http://www.howstuffworks.com; [4] Reprinted with permission from the Privacy Rights Clearinghouse, a non-profit consumer advocacy and information program located in San Diego, CA. ©2002. Contact the Privacy Rights Clearinghouse at 3100-5th Ave., Suite B, San Diego, CA 92103, (619) 298-3396 (voice), (619) 298-5681 (fax), prc@privacyrights.org (E-mail). For the most recent version of these fact sheets, or additional information, visit www.privacyrights.org.

Anonymous Call Rejection (ACR): A companion service to Caller ID, it requires an incoming call from a blocked number to be unblocked before the call will ring through.[4]

B

badge fraud: A type of charity scheme where fraudulent telemarketers purport to be soliciting funds to support police- or fire department-related causes.

bandwidth: The capacity of a telecom line to carry signals. The necessary bandwidth is the amount of spectrum required to transmit the signal without distortion or loss of information. FCC rules require suppression of the signal outside the band to prevent interference.

broadband: Broadband is a descriptive term for evolving digital technologies that provide consumers a signal switched facility offering integrated access to voice, high-speed data service, video-demand services, and interactive delivery services.

bug: A bug is a device placed in an office, home, hotel room, or other area to monitor conversations (or other communications) and transmit them out of that area to a listening post.

C

Call Trace: Immediately after receiving a harassing call, you enter the code *57 on your phone and the call is automatically traced (1157 on rotary phones). Call Trace is easier than using a Trap since the customer does not have to keep a phone log. But Call Trace technology works only within the local service area. Call Trace must be set up in advance by the individual receiving harassing calls, and it requires a fee for use.[4]

Caller ID: Service by which customers who pay a monthly fee and purchase a display device can see the number and name of the person calling before picking up the phone.[4]

calling party pays: A billing method in which a wireless phone caller pays only for making calls and not for receiving them. The standard American billing system requires wireless phone customers to pay for all calls made and received on a wireless phone.

carrier-current technology: Home wireless intercoms are transmitter and receiver sets that use home AC wiring for two purposes. First, they derive power for the units from the AC current. Second, they use the same AC wiring as a form of antenna to transmit and receive the signals from one unit to another. This is called "carrier current" technology.

cellular technology: This term, often used for all wireless phones regardless of the technology they use, derives from cellular base stations that receive and transmit calls. Both cellular and PCS phones use cellular technology.

closed captioning: A service for persons with hearing disabilities that translates television program dialog into written words on the television screen.

commercial leased access: Manner through which independent video producers can access cable capacity for a fee.

common carrier: In the telecommunications arena, the term used to describe a telephone company.

communications assistant: A person who facilitates telephone conversation between text telephone users, users of sign language or individuals with speech disabilities through a Telecommunications Relay Service (TRS). This service allows a person with hearing or speech disabilities to communicate with anyone else via telephone at no additional cost.

Community Antenna Television (CATV): A service through which subscribers pay to have local television stations and additional programs brought into their homes from an antenna via a coaxial cable.

Counterfeit Access Device Law: A federal law which was amended to make it illegal to use a radio scanner "knowingly and with the intent to defraud" to eavesdrop on wire or electronic communication.[4]

cramming: A practice in which customers are billed for enhanced features such as voice mail, caller-ID and call-waiting that they have not ordered.

cross-border telemarketing schemes: Consist of telemarketing schemes—usually advance-fee loan schemes, investment schemes, lottery schemes, and prize-promotion schemes—where the scheme's operators conduct their telemarketing activities in one country and solicit victims in another country.

Customer Proprietary Network Information (CPNI): Your local, long distance or cellular telephone company knows what numbers you call, how often you call them, how much you pay to call them, what services you subscribe to, how you use those services, and other personal and sensitive information about your telephone usage. This information is called Customer Proprietary Network Information (CPNI).

D

diagnostic mode: A cellular telephone can be turned into a microphone and transmitter for the purpose of listening to conversations in the vicinity of the phone. This is done by transmitting to the cell phone a maintenance command on the control channel. This command places the cellular telephone in the "diagnostic mode."

dial around: Long distance services that require consumers to dial a long-distance provider's access code (or "10-10" number) before dialing a long-distance number to bypass or "dial around" the consumer's chosen long-distance carrier in order to get a better rate.

DICTIONARY: A special system of computers used by ECHELON which finds pertinent information by searching for key words, addresses, etc.[1]

Digital Television (DTV): A new technology for transmitting and receiving broadcast television signals. DTV provides clearer resolution and improved sound quality.

Direct Broadcast Satellite (DBS/DISH): A high-powered satellite that transmits or retransmits signals which are intended for direct reception by the public. The signal is transmitted to a small earth station or dish (usually the size of an 18-inch pizza pan) mounted on homes or other buildings.

E

E-911: Enhanced 911.

ECHELON: ECHELON is the term popularly used for an automated global interception and relay system operated by the intelligence agencies in five nations: the United States, the United Kingdom, Canada, Australia and New Zealand (it is believed that ECHELON is the code name for the portion of the system that intercepts satellite-based communications).[1]

e-mail: Also called electronic mail, refers to messages sent over the Internet. E-mail can be sent and received via newer types of wireless phones, but you generally need to have a specific e-mail account.

Electronic Communication Privacy Act of 1986 (ECPA): This law makes it illegal for private citizens to own, manufacture, import, sell or advertise any eavesdropping device while "knowing or having reason to know that the design of such device renders it primarily useful for the purpose of the surreptitious interception of wire, oral, or electronic communications, and that such device or any component therefor has been or will be sent through the mail or transported in interstate or foreign commerce."

enhanced service providers: A for-profit business that offers to transmit voice and data messages and simultaneously adds value to the messages it transmits. Examples include telephone answering services, alarm/security companies and transaction processing companies.

en banc: An informal meeting held by the Commission to hear presentations on specific topics by diverse parties. The Commissioners, or other officials, question presenters and use their comments in considering FCC rules and policies on the subject matter under consideration.

F

Federal Subscriber Line Charge: This is a fee the government allows your local phone company to charge you in order to pay for the telephone lines connected to your home. It isn't a tax that goes to the government, but rather a fee the phone company gets for putting in and maintaining those lines.[3]

footprint: Area in which satellite signals can be received on earth.

Frequency Modulation (FM): A signaling method that varies the carrier frequency in proportion to the amplitude of the modulating signal.

G

Global Positioning System (GPS): A US satellite system that lets those on the ground, on the water or in the air determine their position with extreme accuracy using GPS receivers.

H

handshake: The signal that two fax machines [or computers] will emit to get synchronized to send data.

harassing phone call: When someone calls and uses obscene or threatening language, or even heavy breathing or silence to intimidate you, you are receiving a harassing call.[4]

High Definition Television (HDTV): An improved television system which provides approximately twice the vertical and horizontal resolution of existing television standards. It also provides audio quality approaching that of compact discs.

I

Interactive Video Data Service (IVDS): A communication system, operating over a short distance, that allows nearly instantaneous two-way responses by using a hand-held device at a fixed location. Viewer participation in game shows, distance learning and e-mail on computer networks are examples.

Instructional Television Fixed Service (ITFS): A service provided by one or more fixed microwave stations operated by an educational organization and used to transmit instructional information to fixed locations.

L

landline: Traditional wired phone service.

land mobile service: A public or private radio service providing two-way communication, paging and radio signaling on land.

listening post: This is a secure area where the signals can be monitored, recorded, or retransmitted to another area for processing. The listening post may be as close as the next room or as far as several blocks. Voice-activated equipment is available to record only when activity is present. A recorder can record up to 12 hours of conversation between tape changes.

location tracking: The tracking of a wireless device user's specific location whenever the user's device is on.

Low Power FM radio (LPFM): A broadcast service that permits the licensing of 50–100 watt FM radio stations within a service radius of up to 3.5 miles and 1–10 watt FM radio stations within a service radius of 1 to 2 miles.

Low Power Television (LPTV): A broadcast service that permits program origination, subscription service or both via low powered television translators. LPTV service includes the existing translator service and operates on a secondary basis to regular television stations. Transmitter output is limited to 1,000 watts for normal VHF stations and 100 watts when a VHF operation is on an allocated channel.

M

Municipal Charge: A charge used to pay for local community services such as 911 and other emergency services.[3]

must-carry (retransmission): A 1992 Cable Act term requiring a cable system to carry signals of both commercial and noncommercial television broadcast stations that are "local" to the area served by the cable system.

N

network: Any connection of two or more computers that enables them to communicate. Networks may include transmission devices, servers, cables, routers and satellites. The phone network is the total infrastructure for transmitting phone messages.

number portability: A term used to describe the capability of individuals, businesses and organizations to retain their existing telephone number(s)—and the same quality of service—when switching to a new local service provider.

Number Portability Service Charge: This is an FCC-approved charge that your phone company puts on your bill when you switch long-distance carriers. It pays for the administrative costs of switching from one long-distance carrier to another.[3]

O

off-peak period: The time period from 7 in the evening until 7 in the morning is considered "off-peak" and is charged the lowest rates.[3]

one-in-five scheme: A telemarketing scheme in which the telemarketer contacts a prospective victim and represents that the victim has won one of five (sometimes four or six) valuable prizes.

open video systems: An alternative method to provide cable-like video service to subscribers.

Operator Service Provider (OSP): A common carrier that provides services from public phones, including payphones and those in hotels/motels.

opt-in: A telephone company using this method will not use, or share with its affiliates, the customer's CPNI to market to the customer products and services that the customer does not currently subscribe to, unless the customer expressly gives the company permission to do so. In that way, the customer "opts-in" to the company's use of his or her CPNI.

opt-out: A telephone company using this method sends the customer a notice telling him or her that the company will use (and/or share with its affiliates) his or her CPNI to market products and services that the customer does not currently subscribe to—unless the customer tells the company not to do so. This is known as the "opt-out" method, because the customer's approval is assumed unless he or she "opts-out" of the company's use of the CPNI.

P

paging system: A one-way mobile radio service where a user carries a small, lightweight miniature radio receiver capable of responding to coded signals. These devices, called "pagers," emit an audible signal, vibrate or do both when activated by an incoming message.

peak period: The time period from 7 in the morning until 7 in the evening is considered the "peak" period and is charged the highest rates.[3]

per call blocking: Blocking option in which the phone number is sent to the party being called unless *67 is entered before the number is dialed.[4]

per line blocking: Blocking option in which the phone number will automatically be blocked for each call made from that number.[4]

Personal Communications Service (PCS): Any of several types of wireless, voice and/or data communications systems, typically incorporating digital technology. PCS licenses are most often used to provide services similar to advanced cellular mobile or paging services. However, PCS can also be used to provide other wireless communications services, including services that allow people to place and receive communications while away from their home or office, as well as wireless communications to homes, office buildings and other fixed locations.

pickup device: A microphone, video camera or other device picks up sound or video images and converts them to electrical impulses. If the device can be installed so that it uses electrical power already available in the target room, this eliminates the need for periodic access to the room to replace batteries. Some listening devices can store information digitally and transmit it to a listening post at a predetermined time. Tiny microphones may be coupled with miniature amplifiers that filter out background noise.

predictive dialing: When a telemarketer's automatic dialer simultaneously dials many more numbers than the telemarketer can handle if all of the called parties pick up at the same time, the first to pick up is connected to the telemarketer while the rest are disconnected.

Presubscribed Interexchange Carrier (PIC): Term the telecommunications industry uses to refer to a long-distance carrier.[3]

Prescribed Interexchange Charge (PICC): The charge the local exchange company assesses the long distance company when a consumer picks it as his or her long distance carrier.

pressure valve strategy: Some experts recommend in telephone harassment situations to get a new phone number, but keep the phone number being called by the harasser and attach a voice mail machine or message service to that line. Turn the phone's ringer off and don't use that phone line for anything other than capturing the calls of the harasser.[4]

Privacy Manager: Service that works with Caller ID to identify incoming calls that have no telephone numbers. Calls identified as "anonymous," unavailable," out of area" or "private" must identify themselves in order to complete the call. Before your phone rings, a recorded message instructs the caller to unblock the call, enter a code number, or record their name.[4]

R

Radiofrequency (RF) energy: Another name for radio waves. It is one form of electromagnetic energy that makes up the electromagnetic spectrum. Some of the other forms of energy in the electromagnetic spectrum are gamma rays, x-rays and light. Electromagnetic energy (or electromagnetic radiation) consists of waves of electric and magnetic energy moving together (radiating) through space. The area where these waves are found is called an electromagnetic field.

recovery room: A telemarketer for a "recovery room" contacts the victim, and invariably claims some affiliation with a government organization or agency that is in a position to help telemarketing victims recover some of their past losses.

rip-and-tear scheme: Instead of conducting their telemarketing from a single, fixed place of business, "rip-and-tear" telemarketers conduct their calls from various places, such as pay telephones, residences, and hotel rooms. Their contacts with prospective victims—who usually are repeat victims of past telemarketing schemes—involve explicit promises that the victims have won a valuable prize or are entitled to receive a portion of their past losses. "Rip-and-tear" schemes often insist that the victims send the required "fees" to commercial mailbox facilities or by electronic wire transfer services, which create far less substantial paper trails than checks or credit cards and which allows the telemarketers to receive the victims' payments in cash. To create further difficulties for law enforcement, "rip-and-tear" telemarketers often hire persons to act as couriers to pick up the payments from the mailbox drop or wire transfer office; if law-enforcement agents pick up the payments and arrest the couriers, the organizers of the scheme are not tied directly to the delivery of the victims' funds.

roaming: The use of a wireless phone outside of the "home" service area defined by a service provider. Higher per-minute rates are usually charged for calls made or received while roaming. Long distance rates and a daily access fee may also apply.

roving wiretaps: Wiretaps that are unlike conventional wiretaps in that they allow law enforcement officials to follow the suspect from one location to the next, without having to seek court authorization to wiretap each location's telephone line or other communication channel. In short, the government may wiretap any telephone that the target uses or is known to use.[2]

S

satellite: A radio relay station that orbits the earth. A complete satellite communications system also includes earth stations that communicate with each other via the satellite. The satellite receives a signal transmitted by an originating earth station and retransmits that signal to the destination earth station(s). Satellites are used to transmit telephone, television and data signals originated by common carriers, broadcasters and distributors of cable TV program material.

Satellite Home Viewer Improvement Act of 1999 (SHVIA): An Act modifying the Satellite Home Viewer Act of 1988, SHVIA permits satellite companies to provide local broadcast TV signals to all subscribers who reside in the local TV station's market. SHVIA also permits satellite companies to provide "distant" network broadcast stations to eligible satellite subscribers.

Satellite Master Antenna Television (SMATV): A satellite dish system used to deliver signals to multiple dwelling units (e.g., apartment buildings and trailer parks).

scanner: A radio receiver that moves across a wide range of radio frequencies and allows audiences to listen to any of the frequencies.

service plan: The rate plan you select when choosing a wireless phone service. A service plan typically consists of a monthly base rate for access to the system and a fixed amount of minutes per month.

service provider: A telecommunications provider that owns circuit switching equipment.

slamming: The term used to describe what occurs when a customer's long distance service is switched from one long distance company to another without the customer's permission. Such unauthorized switching violates FCC rules.

SMS spoofing: Transmission of fake text messages.[4]

spectrum: The range of electromagnetic radio frequencies used in the transmission of sound, data and television.

Spread Spectrum Technology (SST): Feature used by some cordless phones that breaks apart the voice signal and spreads it over several channels during transmission, making it difficult to capture.[4]

Subscriber Line Charge (SLC): A monthly fee paid by telephone subscribers that is used to compensate the local telephone company for part of the cost of installation and maintenance of the telephone wire, poles and other facilities that link your home to the telephone network. These wires, poles and other facilities are referred to as the "local loop." The SLC is one component of access charges.

T

tariff: The documents filed by a carrier describing their services and the payments to be charged for such services.

TSCM: Technical Security Countermeasures inspection.

Telecommunications Relay Service (TRS): A free service that enables persons with TTYs, individuals who use sign language and people who have speech disabilities to use telephone services by having a third party transmit and translate the call.

Telephone Consumer Protection Act: Federal law under which it is against the law to use autodialers or prerecorded messages to call numbers assigned to pagers, cellular or other radio common carrier services except in emergencies or when the person called has previously communicated their consent.[4]

telephony: The word used to describe the science of transmitting voice over a telecommunications network.

Third Party Verification (TPV): Telecommunications industry service which verifies that the customer is actually requesting that his or her service be switched to another long-distance carrier.[3]

transmission link: The electrical impulses created by the pickup device must somehow be transmitted to a listening post. This may be done by a radio frequency transmission or by wire. Available wires might include the active telephone line, unused telephone or electrical wire, or ungrounded electrical conduits. Transmitters may be linked to an existing power source or be battery operated. The transmitter may operate continuously or, in more sophisticated operations, be remotely activated.

trap: Service which allows the phone company to determine the telephone number from which harassing calls originate. You must keep a log noting the time and date the harassing calls are received. Traps are usually set up for no more than two weeks. The phone company does not charge a fee for Traps.[4]

TTY ("text telephone" or "teletypewriter"): A type of machine that allows people with hearing or speech disabilities to communicate over the phone using a keyboard and a viewing screen. It is sometimes called a TDD (telecommunications device for the deaf).

turndown room: An operation, affiliated with the telemarketer in an advance-fee loan scheme, whose sole function is to notify the victims at a later date that their loan applications have been rejected.

U

unbundling: The term used to describe the access provided by local exchange carriers so that other service providers can buy or lease portions of its network elements, such as interconnection loops, to serve subscribers.

universal service: The financial mechanism which helps compensate telephone companies or other communications entities for providing access to telecommunications services at reasonable and affordable rates throughout the country, including rural, insular and high costs areas, and to public institutions. Companies, not consumers, are required by law to contribute to this fund. The law does not prohibit companies from passing this charge on to customers.

V

Very High Frequency (VHF): The part of the radio spectrum from 30 to 300 megahertz, which includes TV Channels 2–13, the FM broadcast band and some marine, aviation and land mobile services.

video description: An audio narration for television viewers who are blind or visually disabled, which consists of verbal descriptions of key visual elements in a television program, such as settings and actions not reflected in dialog. Narrations are inserted into the program's natural pauses, and are typically provided through the Secondary Audio Programming channel.

W

war-driving: A new past-time among hobbyists and corporate spies. The data voyeur drives around a neighborhood or office district using a laptop and free software to locate unsecured wireless networks in the vicinity, usually within 100 yards of the source. The laptop captures the data that is transmitted to and from the network's computers and printers.[4]

WECA: Wireless Ethernet Compatibility Alliance, which developed the wireless fidelity (WI-FI) interoperability standard.

WEP: Wired Equivalent Privacy.

WI-FI: Wireless fidelity interoperability standard.

wireless network: Wireless Ethernet networks are built using radio waves. The Institute of Electrical and Electronics Engineers (IEEE) 802.11 standard defines the physical layer and media access control (MAC) layer for wireless local area networks (LANs). As with our wireless telephone networks, the basic building block of the 802.11 architecture is the cell, also known as the Basic Service Set (BSS). A BSS typically contains one or more wireless stations and a central base station. Base stations are the access points to the network and may be either fixed or mobile. All the base stations in a particular wireless network communicate with each other using the IEEE 802.11 wireless MAC protocol.

Wireless Telephone Spam Protection Act: A bill introduced in Congress in 2001, (H.R. 113), would make it illegal to transmit unsolicited ads to wireless devices, including cell phones, pagers, and PDAs enabled to receive wireless e-mail.[4]

Chapter 50

Glossary of Cellular/Wireless Terms

0–9

1G (First Generation Wireless): a term used to describe the first generation of wireless technology (analog cell phones). The systems were designed only to carry voice technology.

1xRTT: the name for the first phase in CDMA's evolution to third-generation (3G) technology. 1xRTT networks allow for increased network capacity (more users; fewer dropped calls), better battery life, and increased data speeds (up to 144Kbps). According to Qualcomm, the developers of the technology, 1x stands for a single radio channel, while RTT stands for radio transmission technology.

2G (also known as (PCS) Personal Communications Services): a term used to describe the second generation of wireless technology (digital cell phones). 2G technology converts voice to digital data for transmission over the air and then back to voice. 2G is the current wireless service available in North America.

2.5G: second-and-a-half generation wireless technology. Most carriers will move to this wireless technology before making the upgrade to 3G. A 2.5G network with GPRS or 1xRTT will change existing wireless

---

This glossary was produced by Branden J. Shortt, Chief Executive, WCS Cellphones Online Inc. © 2002. For additional information, visit www.cell phones.ca. Reprinted with permission. Although there are references to Canadian standards and authorities, readers will find the information in this glossary useful.

networks to a packet-switched service that will increase data transmission speeds.

3-way calling: allows you to conduct a conference call between three parties. (network and subscription dependent feature—not available in all areas)

3G (Third Generation Wireless): a term used to describe the next generation of wireless technology which will provide users with high speed data transmissions (up to 2Mbps) and the ability to roam globally. Known as IMT 2000 by the ITU and implemented in Europe as UMTS and cdma2000 in North America.

3GPP (3rd Generation Partnership Project): a cooperation of standards organizations (ARIB, CWTS, ETSI, T1, TTA and TTC) throughout the world that is developing the technical specifications for third generation wireless technology.

4G (Fourth Generation Wireless): communications systems that are characterized by high-speed data rates at 20+ Mbps, suitable for high-resolution movies and television. Initial deployments are anticipated in 2006–2010.

802.11: refers to a family of specifications for wireless local area networks (WLANs) developed by a working group of the Institute of Electrical and Electronics Engineers (IEEE). There are currently four specifications in the family: 802.11, 802.11a, 802.11b, and 802.11g.

802.11a: refers to a new wireless local area network technology that operates in the 5 gigahertz spectrum. 802.11a is able to transmit data at speeds up to 54 Mbps and helps eliminate interference from devices operating at 2.4 gigahertz, such as cordless phones and microwave ovens.

802.11b: often called Wi-Fi, is the most widely used wireless local area network technology. 802.11b technology operates in the 2.4 GHz range offering data speeds up to 11 megabits per second. A user with a Wi-Fi product can use any brand of access point with any other brand of client hardware that is built to the Wi-Fi standard.

A

AC (Alternating Current): the standard electricity type found in North America.

AC charger: an accessory device that allows you to power and/or charge your phone from a wall outlet.

access point: a base station in a wireless local area network that allows individuals to use wireless networking cards in their computers and other electronic devices. Access points are typically stand-alone devices that plug into an Ethernet hub or server. Depending on the radio environment of the specific building, one access point can provide up to 300 feet (100 meters) of wireless network coverage. Like a cellular phone network, users can roam between access points with their mobile devices and be handed off from one access point to another.

activation: the process by which a cell phone account is created, your phone number assigned, and your phone programmed so that you can make and receive calls.

activation fee: the fee charged by service providers to create an account, assign a phone number and configure a phone with their network.

active flip/keypad cover: a feature that will answer a call by opening the keypad cover and end a call by closing the keypad cover.

Active Matrix Display: see TFT.

Advanced Mobile Phone Service: see AMPS.

aftermarket: a term used to describe an accessory that is made by a company other than the original manufacturer of the product.

air interface: a wireless network's operating system, enabling communication between a cellular phone and its carrier. The main interface technologies are AMPS, TDMA, CDMA, GSM, and iDEN.

air time: the actual time spent using a cellular system. Billing begins when the SEND key is pressed and finishes when the user presses END.

alarm clock: an alarm feature which can be set for a specific time and date or can used as a daily alarm.

alphanumeric display: a display capable of containing both letters ("alphas") and numbers ("numeric").

alphanumeric memory: a special type of dial-from-memory option that displays both the name of individual and that individuals phone number on the cellular phone handset. The name also can be recalled by using the letters on the phone keypad. By contrast, standard memory dial recalls numbers from number-only locations.

AMPS (Advanced Mobile Phone Service): the standard for analog cellular telephones which uses a frequency-modulated transmission and spacing to separate transmissions. Operates in the 800 megahertz (MHZ) band.

AMPS modem: a wireless modem designed for analog cellular phones.

analog: a technology which utilizes a continuous "wave" of signal to carry information over radio channels. In contrast to digital technology, which allows upwards of 15 calls per channel, analog only permits 1 call per channel. Early cell phones all used analog technology. Although analog phones are still common, the majority of new handsets are digital and some carriers no longer offer analog service.

ANSI-41: a protocol standardized by the Telecommunications Industry Association (TIA) and the American National Standards Institute (ANSI) for enabling cdmaOne, cdma2000 and TDMA subscribers to roam between different wireless service operators' systems to make and receive voice calls.

ANSI-136: another name for Time Division Multiple Access (TDMA).

antenna: a part of a cell phone that receives and transmits cellular radio-frequency transmissions.

any-key answer: a feature which enables a user to answer incoming calls easily by pressing any button on the keypad.

any-time minutes: refers to minutes which can be used anytime, without regard to peak/off-peak, day/night, or weekday/weeknight restrictions. Usually a specified number of these minutes are provided with a wireless plan.

ARM: one of the three types of processors that can be found in Pocket PCS. Created by ARM Ltd., the ARM processor has a unique architecture compared to its two competitors (MIPS and SH3), and therefore can only run programs created specifically for it.

asynchronous mode: a transmission data standard, where data information is sent at non-regular intervals. Information is sent as necessary, instead of synchronized with a time signal.

attenuation: the decrease in signal strength as a result of absorption and scattering of energy by objects such as buildings, trees, people, etc.

authentication: a process that allows cellular phones and operators to confirm the identity of any phone that registers itself on the network trough doing or receiving a call.

automatic answer: a feature that allows a user to answer incoming calls without pressing any keys. This feature is generally used in conjunction with a hands-free device.

automatic lock: when activated the phone will automatically lock each time it is power is turned off to help prevent unauthorized use.

automatic redial: automatically redials a busy number simply by pressing the send button.

B

back-lit illumination: illuminates a wireless device's display and keypad for better low light viewing.

bag phone: see Carry/Transportable Phone.

band: a specific range of frequencies in the radio frequency (RF) spectrum.

bandwidth: the amount of data that can be transmitted in a fixed amount of time. Usually expressed in bits per second (bps) or bytes per second for digital devices and cycles per second, or Hertz (Hz) for analog devices.

base station: see Cell Site.

battery capacity: the capacity of wireless devices' battery. Measured in milliampere hours (mAh).

battery indicators: a feature which alerts you that the battery is running low with either an audible tone, or a visual indicator.

battery strength meter: a visual indicator of the estimated time remaining on the battery. Helps avoid dropped calls due to insufficient current voltage.

Bits Per Second: see BPS.

Bluetooth: a wireless personal area network (PAN) specification that connects phones, computers, appliances, etc. over short distances without wires by using low power radio frequencies.

BPS (Bits Per Second): a measure of how fast binary digits can be sent through a channel. The number of 0s and 1s that travel down the channel per second.

BREW (Binary Runtime Environment for Wireless): is an open source application development platform for wireless devices equipped for code division multiple access (CDMA) technology. Developed by Qualcomm, BREW makes it possible for developers to create portable applications that will work on any handsets equipped with CDMA chipsets. A similar and competing platform is J2ME (Java 2 Micro Edition), from Sun Microsystems.

browser: a software application that allows access to the services on the Internet. WAP phones incorporate a special browser that can access mobile Internet services designed specially for the WAP market (characterized by smaller screen sizes and slower data transmission speeds).

built-in charger: a built-in battery charger that allows you to plug the phone directly into a power source to charge any attached batteries.

C

call blocking: allows you to set your phone to prohibit incoming or outgoing phone calls from specific numbers. (network and subscription dependent feature—not available in all areas).

call forwarding: allows users to redirect calls to an alternate telephone number. (network and subscription dependent feature—not available in all areas).

call-in-absence indicator: a feature that, if a phone is left active and an incoming call is not answered, the message "Call" will be displayed to inform the user of a call attempt.

call log: a feature which allows a user to display the numbers of the last incoming and outgoing calls.

call quality: a measure of the total quality of a call including the ability to accurately reproduce a users voice, as well as the systems ability to limit impairments during the course of a conversation.

call restriction: a feature which enables a user to prevent calls to certain numbers without the input of a code.

call timers: enables the tracking of airtime usage to monitor phone expenses. The length of an individual call or a running total (cumulative) can be displayed.

call waiting: a feature that will alert you of another incoming call and enables you to accept the call without disconnecting the first. (network and subscription dependent feature—not available in all areas).

Caller ID: see CLI.

cancellation fee: the cost to terminate the plan prior to the end date specified in the contract

car charger: see CLA.

car kit: a kit that adapts a hand-held cell phone for handsfree use in the car.

car phone: a phone which is permanently installed into a vehicle. They are considerably more powerful (3-watt output) than a handheld cell phone but considerably less flexible.

carrier: a wireless network operator is often referred to as a carrier. Carrier is also a technical radio term for the radio wave that carries voice or data.

carry/transportable phone: a term given to cellular phones that are capable of 3-watt output and can be used either as a portable unit or installed in a vehicle.

CDMA (Code Division Multiple Access): a type of digital wireless technology that allows large amounts of voice and data to be transmitted on the same frequency. CDMA is second-generation cellular technology (or 2G) and is currently available in Canada, the United States, Pacific Asia, and Latin America. Most CDMA service providers (Telus Mobility and Bell Mobility for example) will migrate to a high-speed data technology called 1xRTT.

cdmaOne: the original CDMA (2G) that is in use today in all CDMA networks that have not been upgraded to cdma2000.

cdma2000: defines the third-generation (3G) version of CDMA technology. Also known as IMT-CDMA Multi-Carrier or IS-136, cdma2000 supports high-speed data transmission (144 Kbps to 2 Mbps), always-on data service, and improved voice network capacity (more people can use each tower at the same time). cdma2000 is a competitor to WCDMA and will be deployed in at least three phases—1xRTT, 1xEV-DO, 1xEV-DV, and cdma2000 3x.

Cdma2000 1xEV-DO: the second phase of cdma2000 following 1xRTT deployment. 1xEV-DO stands for 1x Evolution Data Only. "EV-DO" puts voice and data on separate channels in order to provide high-speed, high-capacity wireless Internet connectivity (peak data rate of 2.4 Mbps).

Cdma2000 1x EV-DV: the third phase of cdma2000 following 1xEV-DO deployment. 1xEV-DV stands for 1x Evolution Data Voice, and is characterized by a maximum data rate of 5.2 Mbps and the ability to support wireless Voice over IP (VoIP) services.

CDPD (Cellular Digital Packet Data): an add-on technology that enables first-generation analog systems to provide packet data with a special modem. CDPD modems are available on PC Cards for laptop and handheld computers. CDPD has been implemented in Canada by Telus Mobility in Canada and in the U.S. by AT&T Wireless and Verizon Wireless.

cell: a geographical area of a cellular system in which radio frequency coverage is provided. Also, the basis for the generic industry term "cellular." The cells can vary in size depending upon terrain, capacity demands, etc. but are usually hexagonal and can be anywhere from 0.4 miles up to 15 or more miles in radius.

cell site: a fixed cellular tower and radio antenna that handles communication with subscribers in a particular area or cell. A cellular network is made up of many cell sites, all connected back to the wired phone system.

cell splitting: a means of increasing the capacity of a cellular system by subdividing or splitting cells into two or more smaller cells.

cellular: a wireless telephone service that provides two-way voice and data communications through handheld, portable, and car-mounted phones via geographic areas called cells.

Cellular Digital Packet Data: see CDPD.

cellular signal: the radio waves that carry information between your cellular phone and the cellular system.

CHTML (Compact Hyper Text Markup Language): a subset of HTML designed for small devices, such as smart phones and PDAs. cHTML is essentially a simpler form of HTML designed for small devices with small memory, low power CPUs, limited or no storage capabilities, and small mono-color display screens.

Cigarette Lighter Adapter: see CLA.

circuit switched: a communications method which establishes a dedicated channel and occupies a fixed amount of bandwidth for the duration of the transmission, regardless of whether any data is being transferred.

CLA (Cigarette Lighter Adapter): an adapter which supplies power and/or charges a wireless device from a car's cigarette lighter or a 12 volt supply.

CLI (Calling Line Identification): a feature that allows a phone's display to show you the number and sometimes the name of an incoming caller before you answer. Some carriers allow you to "block" your number when you are sending calls.

cloning: a crime whereby criminals with special equipment capture identity codes from analog phones and create "clone" IDs allowing them to charge calls to your cell phone account. Digital phones cannot be cloned in this way and are also less vulnerable to eavesdropping than analog phones.

CLR (Clear): a key which erases the display.

Code Division Multiple Access: see CDMA.

Compact HTML: see CHTML.

conference calling: a service feature that enables a user to connect with two other numbers for a three-way conversation. Also called three-way calling.

contract: see Service Agreement.

coverage: see Service Area.

cradle: an accessory which holds a wireless device. Cradles may also have the capability to charge batteries.

crosstalk: a disturbance caused by the electric or magnetic fields of one telecommunication signal affecting a signal in an adjacent circuit.

On wireless networks, crosstalk can result in your hearing part of a voice conversation from another circuit.

CRTC (Canadian Radio-television and Telecommunications Commission): regulates Canadian telecommunications service providers.

CTIA: Cellular Telecommunications Industry Association.

CWTA: Canadian Wireless Telecommunications Association.

D

D-AMPS: see TDMA.

data: information that a wireless device can process (numbers, letters and symbols).

data card: allows a fax and data compatible phone to connect to a laptop or a handheld computer. You can then use this combination to access the Internet or send/receive faxes.

data compatible: a wireless feature that enables devices to transmit data either from the handset or via a data card.

data/fax capability: the ability for a cell phone to send and receive fax and data files, access the Internet, and send e-mail when connected to mobile office equipment.

data interface/link: an accessory that allows the connection of wireless devices to computers, fax machines, etc. for data transmission.

data services: enables users to access data, transmit data and communicate with computers and networks. (e-mail, Internet, fax, etc..)

data transmission: the transmission of data between computers or over a telecommunications network.

date and time stamp: a feature that records the exact time and date a message was left.

DCS (Digital Cellular System): a GSM network operating at 1800MHZ. Used by Orange and One 2 One in the UK.

dead spot: an area within a wireless network where service is not available.

dedicated message key: a feature which allows quick access to digital voice and text messages.

desktop charger: a cradle-type device which allows you to charge your phone in an upright position and also lets you charge an additional battery at the same time.

detailed billing: a bill which lists details of usage, including the airtime used, telephone numbers called, and any additional charges.

digital: a method of encoding a transmission that involves translating information (in the case of digital phones the information would be a voice conversation) into a series of 0's and 1's. Digital communications technology offers cleaner calls without the static and distortion that is common with analog phones. The majority of new handsets sold today are digital rather than analog technology.

digital phone: a type of wireless phone which transmits and receives digital signals.

Digital Signal Processing: see DSP.

digital TTY/TDD: enables those who are deaf or hard of hearing to use a special TTY device with digital service. Normally, TTY devices are only compatible with analog service.

dimensions: the size of a device

Direct Connect: a term used by Telus Mobility's Mike service to describe their two-way radio feature which allows a group of users to communicate directly without dialing a phone number.

distinctive ringing: a feature that enables a phone to ring in a special way when calls from a designated list of phone numbers are received (network and subscription dependent feature—not available in all areas).

DragonBall: a series of microprocessors (the brains of a computer) developed by Motorola specifically designed for PDAs, smartphones and Internet appliances.

DSP (Digital Signal Processing): refers to manipulating analog information, such as sound or photographs that has been converted into a digital form to improve accuracy and reliability of digital communications.

DTMF (Dual Tone Multi-Frequency): are tones that your phone transmits to communicate with tone activated phone systems like voice mail or bank by-phone.

dual-band: a wireless phone which is able to operate on both 800MHz and 1900MHz digital networks to send and receive calls; basically, the phone can operate in either digital cellular or PCS frequencies.

dual-battery compatible: allows you to use a main and auxiliary battery on your phone for extended talk times.

dual-mode: a wireless phone which is able to operate on both analog and digital networks to send and receive calls.

dual-NAM: a feature which enables a wireless phone to operate on two separate phone numbers.

Dual-Tone Multi Frequency: see DTMF.

E

ear-to-mouth ratio: the relative positions of the mouth and ear on an adult head. Manufacturers pay particular attention to this ergonomic factor when designing all phones.

early termination fee: see Cancellation Fee.

EDGE (Enhanced Data rates for Global Evolution): a technology being promoted by the TDMA and GSM communities that is capable of both voice and 3G data rates up to 384 Kbps. The standard is based on GSM standard and uses TDMA multiplexing technology.

EFR (Enhanced Full Rate): a feature that allows users with EFR compatible handsets to benefit from significantly better call quality through enhanced digital coding. (network and subscription dependent feature—not available in all areas).

EL (Electro Luminescent): a technology used to produce a very thin display screen, called a flat-panel display, used in some handheld computers.

electronic lock: see Lock.

Electronic Serial Number: see ESN.

e-Mail: the electronic transfer and storage of written messages.

e-mail capability: the ability for a mobile phone or PDA to send and receive e-mail. With a modem and installed or optional third party software, you can send and receive e-mail with most mobile phone and PDAs. E-mail capability, however, is limited by the service or method you use to access the e-mail.

emergency one-touch dialing: a memory location reserved for storing an important number. The number can be accessed and called even if the phone is locked.

EMS (Enhanced Message Service): an extension of SMS that enables the sending of a combination of simple melodies, images, sounds, animations and formatted text as a message to another EMS-compatible phone.

END: a key on a wireless phone which terminates a call.

Enhanced Full Rate: see EFR.

Enhanced Message Service: see EMS.

ERI (Enhanced Roaming Indicator): a feature to indicate whether a mobile phone is on its home system, a partner network, or a foreign (roaming) network. ERI capable handsets, when loaded with the proper software and PRL, will illustrate the home or roam condition using a banner with text on the handset display. While many phones can indicate home vs. roaming via an icon, ERI phones can clearly indicate the third "partner network" status, which may carry a different rate schedule.

Enhanced Roaming Indicator: see ERI.

enhanced services: services available from wireless carriers that provide consumers with value-added telephone services, such as voicemail and call waiting.

ESN (Electronic Serial Number): a unique unchangeable number that is embedded into the phone and is transmitted by the phone as a means of identifying itself within the system.

ETSI: European Telecommunications Standards Institute.

evenings/weekends: a designated time when calling rates are low or free. These times are generally from 7pm to 7am on weekdays and all day Saturday and Sunday.

F

face plate: a front housing or casing on some models of phones that can be detached and replaced with coloured designs.

fascias: see Face Plate.

FCC (Federal Communications Commission): an independent United States government agency charged with regulating interstate and international communications by radio, television, wire, satellite and cable.

FCN (Function Key): a non-numeric key used on certain wireless phones to navigate menus and features.

FOMA (Freedom Of Mobile multimedia Access): the name of NTT DoCoMo's WCDMA service.

free first minute: the first minute of airtime on incoming calls which is not billed.

Freedom Of Mobile Multimedia Access: see FOMA.

frequency: the rate at which a wave alternates, usually measured in Hertz (Hz).

fringe area: the outermost area of a cellular system where signals are weaker.

full duplex: incoming and outgoing audio can occur simultaneously, so user can speak and listen at the same time.

function key: see FCN.

G

GAIT (GSM ANSI-136 Interoperability Team): a technology that enables GSM and TDMA networks to interoperate.

General Packet Radio Service: see GPRS.

GHz (gigahertz): 1 billion hertz in the frequency spectrum.

gigahertz: see GHz.

glass mount: a type of antenna which can be mounted on a window without drilling holes.

Global Positioning System: see GPS.

global roaming: the ability to make and receive calls and send and receive SMS while you travel overseas with your regular cell phone number.

Global System for Mobile Communications: see GSM.

GPRS (General Packet Radio Service): a next generation (2.5G) technology standard for high-speed data transmission over GSM networks. GPRS sends data over packets rather than via circuit switch connections on cellular networks which allows for "always on" wireless data connections and speeds up to 115Kbps.

GPS (Global Positioning System): a system of 24 satellites, computers, and receivers that is able to determine the latitude and longitude of a receiver on Earth. By triangulation of signals from three of the satellites, a receiving unit can pinpoint its current location anywhere on earth to within a few meters.

GSM (Global System for Mobile Communications): a type of digital wireless network which has been widely deployed throughout the world. GSM currently operates using three frequencies: 900MHz, 1800MHz and 1900MHz.. Canada and the United States currently support the 1,900MHz GSM band, while most countries in Europe support either the 900MHz or 1,800MHz bands.

GSM 900: GSM networks operating at 900 MHZ.

GSM 1800: GSM networks operating at 1.8 GHz.

GSM 1900: GSM networks operating at 1.9 GHz (primarily in North America).

H

Handheld Device Markup Language: see HDML.

handheld computer: a portable, handheld computing device that acts as an electronic organizer. Handheld computers are typically used for managing addresses, appointments, to-do lists and notes, but some newer models support wireless Internet access, e-mail, and other interactive applications. Also referred to as PDAs, Handhelds come in two major flavors—Palm and Pocket PC.

hand-off: the transfer of a cellular phone conversation from one cell to another as a phone moves through the service area. It is performed so quickly that callers don't notice.

handset: a mobile or cell phone is often referred to as a handset.

hands-free: a feature that allows users to conduct a conversation without holding the phone.

HDML (Handheld Device Markup Language): a language that allows certain web pages to be presented on cellular telephones and personal digital assistants (PDA) via wireless access.

HDR: see CDMA2000.

headset jack: allows an external headset/microphone accessory to be used with a phone so hands-free conversations can take place.

hearing aid compatible: enables hearing impaired persons to use a wireless device through their t-coil compatible hearing aids. (T-coil must be activated, not compatible with all hearing aids)

hertz: see Hz.

High Speed Circuit Switched Data System: see HSCSD.

high-speed data: support for one of the wireless high-speed data protocols (GPRS, 1xRTT).

home area: the geographic area within which a wireless subscriber can call without incurring roaming or long distance charges. s

home only: a mode that can be selected on a cell phone so that it will only operate within range of your home cellular system.

horn alert: when activated, incoming calls will cause a vehicle's horn to sound or headlights to flash to alert the user to return to the vehicle.

HSCSD (High Speed Circuit Switched Data System): enables the transmission of data over current GSM networks at speeds up to 43.2 kbps. HSCSD enables such high speeds by using multiple channels.

hyperlink: a phrase or word on a WAP page which, once highlighted and selected, links the user to another WAP page.

Hz (hertz): the unit for measuring frequency equal to one cycle per second.

I

incoming call: a call received by a wireless phone.

incomplete call: a call that is not answered or the line is busy. Carriers usually do not charge for incomplete calls.

icons: simple pictures which can be transmitted from one mobile phone to another, along with text using SMS text messaging.

iDEN (Integrated Digital Enhanced Network): a wireless technology by Motorola which combines two-way radio, telephone, text messaging and data. Used by Telus Mobility's Mike service and Nextel. Operates in the 800MHz and 1,500MHz bands using TDMA networks.

illuminated keypad: allows you to view a keypad in low lighting.

i-Mode: a packet based information service for mobile phones and business model developed by Japan's NTT DoCoMo for delivery of Web-type content to wireless handsets.

IM (Instant Messaging): a live chat and e-mail service that enables you to find your friends when they are online and send messages or talk via a private chat room.

IMEI (International Mobile Equipment Identifier): a 15-digit number given to every single mobile phone, typically found behind the battery.

IMSI (International Mobile Subscriber Identity): a unique number for every SIM, used with a key for authentication.

IMT 2000: see 3G.

individual call timer: a timer which displays the duration of the last or current call.

Industry Canada: the Canadian federal Department responsible for the regulation, management and allocation of radio spectrum. Establishes technical requirements for various wireless systems.

infrared: see IrDA.

Instant Messaging: see IM.

Integrated Digital Enhanced Network: see iDEN.

integrated PDA: a phone with built-in PDA functionality. Such phones are also referred to as "smart phones" and contain features such as handwriting recognition, large screens, and contact management software.

International Mobile Equipment Identifier: see IMEI.

International Mobile Subscriber Identity: see IMSI.

International Telecommunication Union: see ITU.

IrDA (infared): allows cell phones, PDAs, and other devices to connect to each other for various purposes. For example, a laptop or PDA can exchange data with a desktop computer or use a printer without a cable connection. IrDA requires line-of-sight transmission like a TV remote control.

IrDA port: a transmitter/receiver for infrared signals

IS-54: first generation TDMA in 1991.

IS-95: first generation CDMA (cdmaONE).

IS-136: second generation TDMA in 1994. Also called "Digital AMPS" or "D AMPS."

ISM band: Industrial, Scientific and Medical band. Unlicensed spectrum typically in the 900MHz, 2.4GHz and 5.7GHz bands. Requires spread spectrum techniques at 1 watt.

iTAP: software developed by Motorola and built into some wireless phones and PDAs that makes typing words on a keypad easier. The competitor to iTAP is T9. See Predictive Text Entry.

ITU (International Telecommunication Union): an organization in Geneva, Switzerland established to promote standardized telecommunications on a worldwide basis.

J

J2ME (Java 2 Micro Edition): is a technology that allows programmers to use the Java programming language and related tools to develop programs for wireless and mobile devices such as cellular phones and personal digital assistants (PDAs). The J2ME platform can be used to implement a wide variety of applications, from wireless games to data portals into the Internet or corporate enterprise databases.

Java 2 Micro Edition: see J2ME.

jog-dial: a single multi-function dial which allows single thumb scrolling up and down through menus and selection of items (by pressing the dial inwards).

K

KBps (Kilobytes Per Second): a measure of bandwidth (the amount of data that can flow in a given time) on a data transmission medium. One thousand bytes per second. About the size of one average e-mail message.

Kbps (Kilobits Per Second): a measure of bandwidth (the amount of data that can flow in a given time) on a data transmission medium. One thousand bits per second.

key: a button on a keyboard.

keypad: the set of buttons on a phone.

keypad lock/key guard: a feature that allows a user to lock the keypad so that it will not respond if pressed.

kilobits: see Kbps.

kilobytes: see KBps.

L

LCD (Liquid crystal display): a type of display used on most cell phones, capable of displaying monochrome characters and some pictures. The LCD has low energy requirements and uses dark segments against a lighter background for easy viewing in all lighting conditions. Color LCD displays use two basic techniques for producing color: Passive matrix is the less expensive of the two technologies. The other technology, called thin film transistor (TFT) or active-matrix, produces color images that are as sharp as traditional CRT displays, but the technology is expensive.

LED (Light emitting diode): a semiconductor device that illuminates when electricity passes through it. Often used as an indicator light, or to spell out words and numbers. LEDs come in many colors, and some LEDs contain multiple elements and are therefore capable of multiple colors. Provides good visibility in direct sunlight and in darkness.

licensing fee: a monthly fee customers must pay to service providers for the right to use the radio frequency spectrum. Mandated by Industry Canada.

lithium ion (LiIon): a type of rechargeable battery for cell phones which is generally lighter weight than earlier battery types, has a relatively longer cycle life, and generally does not suffer from "memory" effect.

lithium polymer: a battery technology similar to lithium ion but allows the battery to be molded to any shape allowing greater flexibility for mobile phone designers.

location services: services that deliver information about the geographic location of mobile telecommunications devices.

lock: a feature that prevents unauthorized use of a phone. When activated the phone will automatically lock each time it is turned off. When turned back on, the phone will prompt the user to enter a unlock code before it will allow a call to be placed. Calls, such as emergency or other specially-programmed numbers, may be placed without entering a lock code.

long-distance: a charge incurred when calling to a telephone number outside your local calling area.

long-distance saver: a feature offered by some carriers designed to help reduce your long-distance charges.

low battery warning: a visual and/or audible indication that the battery is approaching discharge.

M

macrocell: describes a physically large communications coverage area (5–20 km in diameter). Macrocells can hold 60–120 channels (capacity) and can have either high or low power. Macrocells are used primarily to cover large areas that have high traffic.

MAh (Milliampere Hours): a measurement used to describe the energy charge that a battery will hold and how long a device will run before the battery needs recharging. The higher the mAh's, the longer the battery will hold a charge. A milliampere hour (mAh) is 1000th of an ampere hour (Ah).

master clear: changes all non-standard user settings in a mobile phone to standard plus clears all memory locations.

master reset: same as a master clear, but it does not clear all a phone's memory locations and call timers.

megahertz: see MHZ.

melody composer: see Ringtone Composer.

memory dialing: a feature of a cell phone that allows frequently called numbers to be stored for quick dialing by pressing one or two buttons.

memory effect: a battery problem caused by repeated charging before a battery is fully drained. This results in deterioration and prevents batteries from accepting a full charge. It occurs most often in NiCd batteries, is less of a problem with Nickel Hydride batteries and even less with Lithium Ion batteries.

memory locations: a space in an internal phone book where you can store frequently dialed telephone numbers.

memory pause: a pause command that can be entered at the end of a stored number to allow for a system response when using credit card numbers or alternate long distance system ID numbers.

memory protect: prevents accidental overwriting and erasure of existing names and/or numbers in memory.

memory scroll: allows sequential viewing of numbers and/or names stored in memory, starting at a chosen point. A fast and easy means of scanning memory locations.

menu: the list of options that allows you to navigate through a cell phone or handheld computer's functions.

message alert: an indicator that notifies a user of missed voice mail or calls.

message key: a dedicated key on a mobile phone that allows a user to retrieve voicemail or digital messages with the touch of a button.

message waiting indicator: an indicator that notifies a user if he/she has any new unread messages to view. (network and subscription dependent feature—not available in all areas).

messaging: synonymous with text paging, e-mail or short messages received on alphanumeric pagers and other wireless devices. See SMS or IM.

MHz (megahertz): a unit of frequency equal to one million cycles per second (Hertz). Wireless phone communications in Canada and the United States occur in the 800 MHZ and 1900 MHZ bands.

micro-browser: a web browser specialized for a cell phone or a PDA and optimized to run in the low-memory and small-screen environment of a handheld device. This allows a user to access and display specially-formatted Internet content (WAP pages) on the Internet in the wireless markup language (WML). Examples of specially-formatted content include stock reports, news, and sports scores.

microcell: describes a physically small communications coverage area (0.5 + 5 km in diameter) used in densely populated areas where wireless traffic volume is high. The microcell, which is linked to a host macrocell, has low power and a low channel count, making it ideal for high traffic city neighborhoods.

MIDI (Musical Instrument Digital Interface): a standard that allows digital musical instruments to communicate with one another. In cell phone terms, MIDI is what gives you polyphonic sounds; which means your ring tones can sound like real music instead of beeps.

MIN (Mobile Identification Number): a 24-bit number assigned by a wireless service provider (carrier) to each phone it sells or includes in a service plan that uniquely identifies a mobile device within a carrier's network. Unlike an Electronic Serial Number (ESN), a MIN is changeable because wireless phones may change hands or phone owners may move to another coverage region, requiring a different service plan.

mini-browser: see Micro-browser.

minutes included: the number of free air time or usage included each month in a cell phone rate plan.

MIPS: one of the three types of processors that can be found in Pocket PCS. Created by MIPS Technologies, the MIPS processor has a unique architecture compared to its two competitors (ARM and SH3), and therefore can only run programs created specifically for it.

missed call indicator: a feature which notifies a caller that a call was received but not answered.

MMM (Mobile Media Mode): an icon that identifies web content optimized for smart phones and handhelds.

MMS (Multimedia Messaging Service): a further extension of SMS and EMS. MMS is designed to make use of newer and quicker mobile transmission methods such as GPRS, HSCSD, EDGE and UMTS, involving the attachment of multimedia extensions to messages, such as video and sound. An e-mail function is also planned.

MO-SMS (Mobile-Originated Short Message Service): the ability to send short text messages from a phone. Both the phone and the carrier's network must support this feature for it to work. Messages can be sent to other phones by phone number. Many phones also allow sending messages directly to e-mail addresses.

mobile commerce: the use of radio-based wireless devices such as cell phones and personal digital assistants to conduct business-to-business and business-to-consumer transactions over wired, Web-based e-commerce systems.

mobile data: a service which enables users to access data, transmit data and communicate with computers and networks. (e-mail, Internet, fax, etc..)

Mobile Identification Number: see MIN.

mobile Internet: access to specially designed Internet sites offering services such as news, travel, weather and entertainment using a wireless application protocol (WAP) phone.

mobile IP: a protocol developed by the Internet Engineering Task Force to enable users to roam to parts of the network associated with a different IP address than what's loaded in the user's appliance.

Mobile Media Mode: see MMM.

mobile phone: a wireless phone or cell phone is often referred to as a mobile phone. Initially, a mobile phone referred to a phone attached to a vehicle, which used the vehicle's battery and had an external antenna.

Mobile Telephone Switching Office: see MTSO.

Mobile Virtual Network Operator: see MVNO.

modem: a device which converts digital data to analog data (tones) so that it can be sent over regular phone lines and wireless networks. The modem also converts data back from analog to digital.

MP3 playback: some cell phones feature a MP3 player (built-in or add-on accessory) that allow you to listen to music stored in the MP3 digital format. These files are much smaller than other formats such as wave files, yet can deliver CD quality sound. Generally, music can be downloaded into the phone from a computer and played back later through a headset attached to the phone. Newer phones with High-Speed Data may support downloading music directly over the Wireless Internet.

MTSO (Mobile Telephone Switching Office): the central switch that controls the entire operation of a cellular system. It is a sophisticated computer that monitors all cellular calls, keeps track of the location of all cellular equipped vehicles traveling in the system, interconnects calls with the local and long distance land line telephone companies, arranges hand-offs, keeps track of billing information etc. Every cellular system has one or more MTSOs or switches.

multi-language display: a feature that allows you to select in which language (English, French, or Spanish) the phone will display messages and prompts.

multi-mode: a wireless device that can operate on either an analog or digital wireless network, allowing you to maintain a connection whether you're in a digital service area or analog only service area.

Multimedia Messaging Service: see MMS.

multiple key answer: a feature that allows you to answer an incoming call by pressing any key. A faster, more convenient way to answer than searching for a specific key.

multiple NAM: a feature which allows a wireless phone to operate on multiple phone numbers and establish accounts with service providers in more than one service area.

multiple numbers per name: allows a user to enter more than one phone number (Home, Cell, Office, Fax) in a single phone book entry.

mute: mutes the handset or speaker to allow private conversations without the called party overhearing.

MVNO (Mobile Virtual Network Operator): functions as a wireless service operator in the marketplace though it does not own an actual wireless network. An examples of a MVNOs is Virgin Mobile.

N

NAM (Number Assignment Module): a circuit chip located inside a phone which stores your telephone number, lock code, timer reset code, network information and other operational data. The NAM is programmed by the service provider when a device is activated. Today's phones have EPROM type NAM and are keypad programmable.

NAMPS (Narrowband Advanced Mobile Phone Service): is the next generation of AMPS systems. NAMPS is a cellular call-handling system that uses digital signaling techniques to split the existing channels into three narrowband channels. The result is three times more voice channel capacity than the traditional AMPS system provides.

Narrowband Advanced Mobile Phone Service: see NAMPS.

net mode: represents the Internet browser mode.

network(s): the companies that supply the transmitters and framework allowing calls to be made in. There are four major nation-wide networks in Canada: Bell Mobility, Microcell (Fido), Rogers AT&T, and Telus Mobility.

NiCd (nickel cadmium): an older type and the most basic type of rechargeable battery technology for cell phones which can be damaged if it is not fully drained before recharging (referred to as memory effect).

nickel cadmium: see NiCd.

nickel metal hydride: see NiMH.

NiMH (nickel metal hydride): a newer type and common from of rechargeable battery for cell phones which will is less sensitive to the memory effect.

no answer/busy transfer: forwards incoming calls to another number when your line is busy or cannot be answered.

no SVC: an indicator on a cell phone which notifies you when you are out of a service or coverage area.

noise canceling microphone: a type of microphone technology that screens out unwanted background noise to allow clearer conversations.

NTT DoCoMo: DoCoMo (meaning "anywhere" in Japanese) is an NTT subsidiary and Japan's biggest mobile service provider. NTT DoCoMo is the chief developer of I-Mode.

Number Assignment Module: see NAM.

numeric messaging/paging: allows you to receive numerical messages.

O

off-peak: a designated time of the day or week when calling rates are cheaper or free. These times are usually in the evenings from 7 pm to 7 am on weekdays or on weekends.

OLED (Organic Light-Emitting Diode): a next-generation display technology that consists of small dots of organic polymer that emit light when charged with electricity. OLED is beginning to replace LCD technology in handheld devices such as PDAs and cell phones because the technology is thinner, lighter, brighter, cheaper to manufacture and consumes less power than LEDs.

on-hook call processing: allows the user to leave a cellphone in its mounting cradle until the called party answers for safer operation.

one-touch emergency dialing: a memory location reserved for storing an emergency number. This feature allows you to connect to an emergency number by pressing a single button and can be accessed and called even if the phone is locked.

operating frequency: the rate at which an electrical current alternates, usually measured in Hertz (Hz).

Organic Light-Emitting Diode: see OLED.

OTA (Over The Air): the downloading of ring tones, picture messages, and other content to your mobile phone wirelessly.

P

P-Java (Personal Java): a Java API and specification for running Java applications on small devices.

packet: a piece of data transmitted over a packet-switching network such as the Internet or wireless Internet; a packet includes not just data but also its destination.

packet switching: a type of communication that splits information into "packets" of data for transmission. This is efficient, as it only uses radio spectrum when it's actually sending something, rather than keeping an open channel at all times (as is done in circuit switching). Packet switching is a core component to 3G technology.

packet-switched network: networks that transfer packets of data (see Packet). These networks are a more reliable method of transferring wireless data than a circuit-switched network. Packet-switched networks eliminate the need to dial in to send or receive information because they are "always on," transferring data without the need to dial.

pager: a one-way or two-way radio receiver device that allows reception and display of a numeric or alphanumeric message. Most new cell phones have similar functionality built-in.

Palm: a handheld computer or PDA that runs the Palm operating system. The Palm operating system which was originally created for Palm PDAs, has since become the OS of choice on PDAs from many different companies like Sony, Kyocera, and Handspring. It features a wide range of organizer functions such as telephone book, e-mail, to-do lists, spreadsheets, word processors, and wireless Internet capabilities. Palm PDAs can usually synchronize with PCS or Apple computers using infrared, Bluetooth or wire connections. Many mobile phones can connect to PDAs using a special connector and an infrared connection to send and receive e-mails. The latest smartphones combine many of these features in a single unit. The main alternative to the Palm is the Pocket PC by Microsoft.

Passive Matrix Display: an LCD technology that uses a grid to supply the charge to each particular pixel on the display. An STN screen has a slower refresh rate than a TFT screen, but it's cheaper. Also called a SuperTwist Nematic of STN display.

pause dialing: a command that can be entered into stored numbers. By including pauses between memory locations the phone can dial a telephone number and then wait for a response before continuing to transmit. This feature is useful for accessing voice mail system, banking via phone, accessing credit card information, etc..

PC card (PCMCIA): a removable, credit-card sized devices that may be plugged into slots in PCS and wireless communication devices to provide fax or modem functions or network cards.

PC sync: allows a user to connect a cell phone to a computer with a cable and transfer data. An example of this would be synchronizing a cell phone's contact and calendar information with a computer application like Outlook.

PCMCIA (Personal Computer Memory Card International Association): a group of hardware manufacturers and vendors responsible for developing standards for PC Cards (also called PCMCIA cards.)

PCN: also known as DCS 1800 or GSM 1800, PCN is a term used to describe a wireless communication technology in Europe and Asia.

PCS (Personal Communications Services): a term used to describe two-way, 1900MHz digital wireless technology. PCS, a second-generation technology, arrived in 1990 and is the most widely deployed wireless service in North America today. It is based on circuit-switched technology where each call requires its own cell channel, which makes transmission of data quite slow. 2G PCS services include Code Division Multiple Access(CDMA), Time Division Multiple Access (TDMA), and GSM.

PDA (personal digital assistant): a portable, handheld computing device that acts as an electronic organizer. PDAs are typically used for managing addresses, appointments, to-do lists and notes, but some newer models support wireless Internet access, e-mail, and other interactive applications. Also referred to as Handheld Computers. PDAs come in two major flavors—Palm and Pocket PC.

PDC (Personal Digital Communications): the digital cell phone system in Japan.

peak minutes/period: a designated time of the day or week when cellular calling rates are highest. These times are generally between 7am and 7pm on weekdays.

Personal Communications Service: see PCS.

personal hands-free kit: a device that allows you to use your phone hands-free by wearing a headset and microphone, rather than holding the phone to your ear.

Personal Identification Number: see PIN.

Personal Information Manager: see PIM.

PHS (Personal Handyphone System): a digital cell phone technology based on TDMA and used in Japan.

phone book: a feature that enables you to store a collection of telephone numbers and names into your phone's internal memory or on its SIM card. Storing numbers in the phone book makes frequent calls easier.

phone lock: a feature which prevents unauthorized use of a phone.

phone only mode: a feature on multi-service phones like iDEN that allow a user to disable the two-way radio mode to increase stand-by time.

photo ID: allows a user to set custom graphics (can be pictures) with a phone book entry. When the person who is associated with that phone book entry calls, the corresponding graphic is shown. Graphics can be downloaded into the phone from a computer, or via the wireless Internet.

picocell: describes a physically small communications coverage area (less than 0.5 km in diameter).

piconet: a network of devices connected using Bluetooth wireless technology. A piconet may consist of two to eight devices. In a piconet, there will always be one master while the others are slaves.

picture messaging: a technology that allows you to send and receive picture messages as well as text on a mobile phone.

PIM (Personal Information Manager): a type of software application that allows the user to input and organize various types of information.

Common features of a PIM application include a notepad, calculator, to-do list, calendar and scheduling tool.

PIN (Personal Identification Number): a numeric code or password that may be required by a service provider in order to make outgoing calls or obtain access to certain applications and data. This code is always associated to a SIM card, not a phone and is designed to help guard against cellular fraud.

plan: see Service Plan.

Pocket PC: a handheld Windows-based computer or PDA that runs the Pocket PC operating system (formerly Windows CE) by Microsoft. The Pocket PC operating system features Pocket Office applications (Internet Explorer, Word and Excel), handwriting recognition, an e-book reader, and wireless Internet capability. The main alternative to the Pocket PC is the Palm OS.

polyphonic ring tones: ring tones very much like regular ring tones except that they are capable of playing multiple notes at a time. This results in vastly improved sound quality with richer, more realistic sounds. Phones equipped with polyphonic ring tones generally have better sounding speakers.

predictive text input: software built into some cell phones and mobile devices that makes typing words on a keypad easier. Instead of pressing each key one, two or three times, just to press it once and a built-in vocabulary will attempt to guess the word that you are spelling. Using this system, SMS messages and sometimes e-mails are quicker and easier to write. Often referred to as T9, the most popular type of predictive text entry. The competitor to T9 is iTAP by Motorola.

preferred system: a cell phone's home system.

pre-paid card: a card or voucher that represents advanced payment for wireless service.

pre-pay/pay as you go: a system allowing subscribers to pay for wireless service usage in advance. There is no activation charge and instead of being billed for your calls, you simply buy a top-up card or voucher that pays for the calls in advance. Prepaid is generally used for credit-impaired customers, those who want to adhere to a budget, or those who do not want to sign a contract. Each network (Bell, Fido, Rogers, Telus, etc.) has its own pre-paid service.

processor speed/CPU: the measure of the speed of the microprocessor of the handheld/PDA in megahertz (MHZ). In general, the higher this number gets the faster the handheld/PDA will execute tasks.

profiles: a group of phone settings (ringing tones, keypad tones, warning tones) that you can customize. With profiles you can create sets of combined tones and screensavers to suit different environments or times of day.

promotion: a special offer on cell phones or services. Promotions usually include a discount on a phone, extra minutes, or a lower monthly access charge.

PTT (Push-To-Talk): a two-way communication service that works like a "walkie talkie". This feature, found on Motorola iDEN phones from Nextel and Telus Mobility's Mike, allow communication in only direction at a time unlike a cell phone that allows for simultaneous conversations. New PTT systems are now being introduced that use VoIP technology to provide PTT service digitally over 3G data networks. See VoIP.

PUK (Personal Unblocking Code): used to unblock a blocked SIM card, this code is given during the subscription of a phone.

Push-To-Talk: see PTT.

PWR: represents the on/off (power) key on some wireless devices.

R

R-UIM (Removable User Identity Module): introduced by the CDMA Development Group, the R-UIM is similar to a SIM card but designed for use with CDMA based mobile phones.

radiation: a radio frequency (RF) radiation emitted by cell phones or other wireless devices, which some studies have suggested may have implications for safe mobile phone use.

rapid charger: a cell phone battery charging accessory that is capable of fully charging a battery in less than four hours.

rate per additional minute: the fee for additional minutes that exceed a rate plan's base minute allotment.

rate plan: see Service Plan.

ringer ID: allows a user to set custom ringtones with a phone book entry. When the person who is associated with that phone book entry calls, the corresponding ringtone will sound. Also called distinctive ringing or name ringer.

ringer options: the ways in which a phone will notify a user of an incoming call. Most phones feature multiple ring tones, melodies, a silent ringer, and a vibration alert. Some handsets now have built-in ringtone composers and some can download ringtones using PC synchronization software or via the wireless Internet.

ringer profiles: allow a user to create distinct "profiles", each consisting of an array of detailed ringer settings. These profiles can be pre-set and stored in the phone by the user, then quickly selected and activated at any time.

ringtone composer: software that allows a user to create their own ringtones by pressing key sequences on a cell phone or using external PC composer software.

ringtones: audible alerts on a cell phone or wireless communication device that notify a user of an incoming call. Phones generally have a small collection of ringtones built-in. Some handsets now have built-in ringtone composers and some can download ringtones using PC synchronization software or via the wireless Internet. The latest ring tones, polyphonic, are capable of playing multiple notes at a time which results in richer, more realistic sounds.

roaming: a service offered by most cellular service providers that allows subscribers to use cellular service while traveling outside their home service area. The areas / countries you can roam in and the cost will depend on which service provider you use. Roaming requires an agreement between operators of technologically compatible systems in individual markets to permit customers of either operator to access the other's systems. (If you require a phone that will operate in Europe, you will need a Tri-band phone).

S

SAR (Specific Absorption Rate): the unit of measurement for the amount of radio frequency (RF) absorbed by the body when using a wireless phone. Usually SAR is expressed in watts per kilogram (W/kg) or milliwatts per kilogram (mW/kg). In the United States the maximum allowable SAR is 1.6 w/Kg. In Europe that value is 2 w/Kg.

satellite phone: a phone that connect callers via satellite. Satellite phones give users a worldwide alternative to sometimes un-reliable digital and analog connections but the systems are costly.

scratch pad: allows you to enter information into a phone's keypad during a conversation without interrupting the call.

screensaver: a picture or animation which appears on a cell phone's display when it is idle.

scroll keys: a key or keys on a mobile phone's keypad that allows a user to scroll forward and backward through menu options and lists.

SDK: a Software Development Kit for wireless application developers.

SDMA (Space Division Multiple Access): a variation of TDMA and CDMA that potentially will be used in high-bandwidth, third-generation wireless products.

security code: a numeric code (password) used to prevent unauthorized or accidental alteration of data programmed into wireless phones. The security code can be used by the owner of a phone to change the lock code.

service agreement: a business contract or agreement with a service provider to use its service for a period of time. The contract typically outlines the services provided and the costs of the services including a monthly base rate (with included minutes) and per-minute charges for minutes over the monthly maximum. In return for your commitment, a service provider will generally subsidize the initial cost of a cell phone.

service area: refers to the geographic area served by a wireless carrier, within which you can use your wireless device to send and receive calls or information. Service areas vary greatly from carrier to carrier. Often used to describe the strength of a service provider's signal. Also referred to as coverage area.

service plan: an agreement with a service provider which gives access to the wireless network, gives an allotted number of minutes per month, and may include features such as call forwarding, call display, etc.

SH3: one of the three types of processors that can be found in Pocket PCS. Created by Hitachi, the SH3 has a unique architecture compared to its two competitors (ARM and MIPS), and therefore can only run programs created specifically for it.

Short Message Service: see SMS.

signal strength meter: a visual indicator which displays the relative strength of the cellular signal to help ensure that quality calls can be placed.

silent keypad: a feature that turns off the tones made by a cell phone when pressing a key.

silent ringer: a feature that signals incoming calls by flashing an indicator light rather than ringing.

silent scratch pad: a feature that silences the tones heard when pressing a key on a cell phone.

SIM (Subscriber Identification Module): a removable plastic card found in GSM phones that stores pertinent information about a phone such as your phone number, account information, phone book, PIM data, etc.. The card can be plugged into any GSM compatible phone and the phone is instantly personalized to the user. SIMs come in two sizes: large (credit card size) and small (thumbnail size).

SIM Lock: software within a phone that can be enabled so that the phone will only work with a one nominated SIM card. Carriers usually block mobile phones to assure they are only used in their network.

slave units: any unit within a Piconet that is not the master unit.

sleep mode: allows the user to conserve battery power when the phone is waiting for a call.

smartphone: a term typically used to describe a next-generation device that combines the functionality of a mobile phone with the enhanced features found in a PDA. Functions such as calendar, telephone book, e-mail, to-do lists, spreadsheets, word processors, and wireless Internet access are typical.

SMR (specialized mobile radio): a dispatch radio and interconnect service for businesses. Covers frequencies in the 220 MHZ, 800 MHZ and 900 MHZ bands.

SMS (Short Message Service): a service that enables subscribers to send short text messages (usually about 160 characters) to and from wireless handsets. These messages can be sent from a Web site or from one wireless phone to another and enhancements are being made to support rich text and graphics. See MMS and EMS. Also called Text Messaging. (network and subscription dependent feature—not available in all areas)

SMS Chat: a feature available on some newer phones that allow a user to "chat" with other users via the sms protocol.

SND (Send): a key on cell phone that initiates the call typed on the keypad and answers incoming calls.

soft handoff: a procedure in which two base stations—one in the cell site where the phone is located and the other in the cell site to which the conversation is being passed—both hold onto the call until the handoff is completed. The first cell site does not cut off the conversation until it receives information that the second is maintaining the call.

speakerphone: enables conversation to take place hands-free. Enhances safety and convenience and can be used to conduct conference calls. There are two distinct types of two-way speaker-phone functionality: Half-Duplex and Full-Duplex. Half-Duplex allows only one person to speak at a time. When one person is speaking, the other person can not be heard at all until the first person has stopped speaking completely. With Full-Duplex, both parties can speak naturally and be heard at the same time, just like non-speakerphone usage.

spectrum: refers to a band of frequencies where wireless signals travel carrying voice and data information.

speed dialing: a feature which allows a user to connect to a phone number by pressing one, two or three digits instead of dialing in an entire phone number.

standby time: refers to the amount of time a battery lasts when a wireless device is turned on but is not in use. When the phone is switched on and waiting for a call it is on 'standby'. A cell phone will consume battery power when on standby but far less than when talking on the phone.

subscribers: the user of an individual handset. In some cases, a client or customer equates to a subscriber, in other cases one client includes multiple subscribers.

Subscriber Identification Module: see SIM.

subsidy: when a cell phone is purchased, it is generally subsidized by whichever network you connect to. You may only pay $1 for a phone worth $300. If the phone is then lost or stolen, the replacement cost will be a higher, unsubsidized price.

switch-hook operation: handles call transferring, three-way calling, and other services the service provider may offer.

stylus: a pen like device usually used coupled with handwriting recognition software for writing on a PDA or mobile phone display. It also works to navigate trough sensitive menus.

Symbian: the name given to a venture formed by Nokia, Ericsson, Motorola, and Psion to create easy to use operating systems for wireless devices and personal digital assistants (PDAs). The first operating system is called EPOC.

synchronous mode: a standard for data communication in which information is transmitted without start or stop bits, along with a clock icon that synchronizes the sender and the receiver. This mode allows bigger transfer rates but it may be more uncertain due to the need of synchronization.

SyncML: an open data synchronization protocol enabling data synchronization between mobile devices and networked services. SyncML is a transport, data type, and platform independent technology that is based on Extensible Markup Language (XML).

T

T9: software built into some wireless phones and PDAs that makes typing words on a keypad easier. The competitor to T9 is iTAP. See Predictive Text Entry.

TACS (Total Access Communications System): a cell phone system in Europe based on analog (AMPS).

talk time: the amount of time a battery lasts when a wireless device is actively transmitting or receiving a call. Talking on your phone uses battery life much quicker than if the phone is on standby.

tariff: the price which you pay for using your phone. This will normally include a monthly fee, an allotted number of minutes per month, and a rate for calls per minute, which may vary by time of day. Each network operates a range of tariffs designed to meet the needs of different customers. Also see Service Plans.

TDMA (Time Division Multiple Access): a family of second-generation digital wireless technologies (GSM, TDMA, iDEN, PDC and PHS) that divides conversations into packets of data according to time. This allows large amounts of voice and data to be transmitted on the same frequency. TDMA runs on two bands: 800MHz and 1,900MHz. TDMA networks are used in North, Central, and South America. TDMA and GSM networks are similar in that they can both share the same migration path to high-speed data: GPRS (2.5G), then EDGE (3G). Also referred to as D-AMPS.

telematics: the integration of wireless communications, vehicle monitoring systems and location devices.

terminal: a device capable of sending, receiving, or sending and receiving information over a communications channel. Also referred to as a mobile terminal, mobile station, or wireless terminal.

termination charges: fees that wireless service providers pay to complete calls on wireline phone networks or vice versa.

text messaging: see SMS.

TFD (Thin Film Diode): a type of LCD (Liquid Crystal Display) flat-panel display technology. TFD technology combines the excellent image quality and fast response times of TFT, with the low power consumption and low cost of STN.

TFT (Thin Film Transistor): an LCD technology that uses transistors to precisely control the voltage to each liquid crystal cell. This is also referred to as an "active matrix" display. TFT screens offer the best image quality and refresh rates, but at a higher cost.

Thin Film Diode: see TFD.

Third Generation Wireless: see 3G.

Time Division Multiple Access: see TDMA.

transportable phone: see Carry Phone.

tri-band: a phone capable of operating on three different digital frequencies—900MHz, 1800MHz and 1900MHz. A Tri-Band phone is able to use networks in Europe, Asia (900MHz, 1800MHz) and North America (1900MHz).

tri-mode: a wireless phone that can operate on both the 1900 and 800MHz digital networks, and on the 800MHz analog network.

triangulation: a lengthy process of pinning down a caller's location using radio receivers, a compass and a map.

trickle charge: a technology used on a battery charger that detects when the battery is fully charged. At that time the charger changes to trickle charge mode assuring that the battery is fully charged. A battery charged in a charger without this technology gradually loses its capacity.

trunking: a spectrum-efficient technology that establishes a queue to handle demand for voice or data channels.

two-way radio: a feature offered by the wireless carrier Mike. This is similar to a walkie-talkie.

theft alarm: a feature on some phones that can be activated to make a call to a pre-programmed number if the unit is not unlocked within a specified time.

time and date stamp: a feature that automatically displays the time and date of an incoming message.

time slot: a unit derived from TDMA technology. There are eight time slots per carrier frequency or TDMA frame. Each can be used for one-way GSM voice or data traffic. A conversation requires two time slots, while HSCSD allows several to be joined together to increase data transfer rates.

tone dialing from keypad: allows calling of numbers requiring tone dialing, just like a conventional phone.

tone dialing from memory: allows storage and recalling of frequently used tone numbers.

top-up vouchers: see Pre-Paid Card.

tower: see Cell Site.

transceiver: a radio transmitter and receiver combined into a single unit.

TTY/TDD ("text telephone" or "teletypewriter"/Telecommunication Devices for the Deaf): terminals used for two-way text conversation over a phone line. They are the primary tool used by deaf people (and some hard of hearing people) for telephone conversation. Other visual telecommunications technologies and services, such as Internet chat and messaging, e-mail, paging, and fax and e-mail are also widely used in telecommunications by people who are deaf or hard of hearing.

two system registration: allows a user to register a phone for operation on two different cellular systems (carriers) with two separate numbers for travel convenience.

two way paging: the ability to receive and send data to the Internet by way of the paging network; also often called interactive paging.

two way radio: a radio transmitter and receiver. Ideal for skiing and active pursuits. They have a range of up to 3km with no call charges and no reliance on network coverage. Also, sometimes used to describe the push-to-talk radio feature available on iDEN cell phones from Nextel and Telus Mobility's Mike. This service is similar to a walkie-talkie but with extended range throughout your calling area. See iDEN.

U

UMTS (Universal Mobile Telecommunications System): a third-generation wireless communications technology and the next generation of GSM (Global System for Mobile Communications). UMTS is a wireless standard approved by the International Telecommunications Union (ITU) and is intended for advanced wireless communications. UMTS promises high-speed mobile data (up to 2 Mbps) and advanced multimedia capabilities such as streaming video.

Universal Mobile Telecommunications System: see UMTS.

unlock code: the digits you enter to unlock a wireless device.

upgrade: some service providers will allow you to upgrade your phone to a newer model after a period of time, this usually involves signing a contract again.

URL (Uniform Resource Locator): a unique name or number that specifies the location of a file on the Internet. A URL consists of a protocol, such as http:// that specifies a web page, followed by a server or path name. For example, the URL for the Cellphones.ca web site is http://www.cellphones.ca

USB (Universal Serial Bus): a plug-and-play interface between a computer and add-on devices (such as keyboards, phones and PDAs). With USB, a new device can be added to a computer without having to add an adapter card or even having to turn the computer off. USB supports a data speed of 12 megabits per second and is now being incorporated in some cell phones which is useful for synchronizing information with a computer or downloading ringtones.

V

Vacuum Fluorescent Display: see VFD.

VFD (Vacuum Fluorescent Display): a type of display used on some cell phones. This display retains visibility in direct sunlight and is highly visible in darkness. It can be seen without distortion over a wide range of viewing angles and remains fully operational over a broad temperature range.

vibration alert: a feature that notifies you of an incoming call or message by vibrating rather than ringing.

vocoder: a device that encodes and decodes the sound of human voice into/from digital format for transmission.

voice activated dialing: a feature that allows a user to dial a phone number by spoken commands instead of punching the numbers in physically. The feature contributes to convenience as well as safe driving.

voice mail: a service that answers calls and records incoming voice messages. Basically, an answering machine on your cell phone. This will take messages if your phone is switched off or you are engaged.

voice mail indicator: a feature that notifies you of messages in your voice mail box.

voice mail key: a key on the keypad of a cell phone or other communications device that allows you to retrieve voicemail or digital messages with the touch of a button.

voice memo: a feature that allows you to record and store short voice messages that you can play back at any time. Many phones with this feature also let you record parts of phone conversations in progress. Some phones have a dedicated voice-memo button to activate the feature.

voice recognition: the capability to control or control certain functions on cell phones and other communications devices by using voice commands.

VoIP (Voice over Internet Protocol): a technology for transmitting voice, such as ordinary telephone calls, over the Internet using packet-switched networks. Also called IP telephony.

volume control: adjusts volume levels on a cell phone or mobile device for the earpiece, ringer, and speaker to personal preference.

VOX (Voice Operated Transmitter): a battery-saving feature that transmits only when talking is taking place.

W

W3C (World Wide Web Consortium): an international industry consortium founded in 1994 to develop common standards for the World Wide Web. It is hosted in the U.S. by the Laboratory for Computer Science at MIT.

W-CDMA (Wideband Code Division Multiple Access): a third-generation (3G) wireless technology that supports high-speed data transmission (144 Kbps to 2 Mbps), always on data service, and improved network capacity (more people can use each tower at the same time) in GSM systems by using CDMA instead of TDMA. The version of WCDMA used by NTT

DoCoMo in Japan is called FOMA or J-WCDMA; the European version is referred to as UMTS, E-WCDMA, or MT-2000 Direct Spread. W-CDMA is a competitor to cdma2000.

wallet: a cell phone software application that enables users to make Internet type payments via a WAP-browser, where card information is transferred from the customer to the Internet merchant. The application is capable of storing protected personal information inside the phone.

wallpaper: a background design on the screen of a cell phone or other mobile device. Some phones allow you to change the design of the wallpaper in much the same way as PC users can do.

WAN (Wide Area Network): a physical or logical network that provides data communications to a larger number of users than are usually served by a local area network (LAN) and is usually spread over a larger geographic area than that of a LAN.

WAP (Wireless Application Protocol): a set of standards that enables a wireless phone or other mobile device to browse Internet content optimized for wireless phones. The competitive technology to WAP is I-Mode by Japan's NTT DoCoMo.

WAP gateway: software that takes raw WML data and compiles it for a micro-browser and vice versa.

WASP (Wireless Application Service Provider): vendors that provide hosted wireless applications so that companies will not have to build their own sophisticated wireless infrastructures.

WBMP (Wireless Bitmap): a bitmap graphic format for integration of images in WAP pages. WBMP graphics are only black and white and have a 1 Bit size.

WCS (Wireless Communications Services): services used to conduct communications over wireless networks.

web clipping: an application that allows a user to extract relevant information from a web page for display on a smart phone or a PDA.

weekends/evenings: a designated time when cellular calling rates are lowest or free. These times are generally between 7pm and 7am on weekdays and all day Saturday and Sunday.

Wide Area Network: see WAN.

Wideband Code Division Multiple Access: see W-CDMA.

Wi-Fi (Wireless Fidelity): the popular term for the 802.11b wireless Ethernet standard. See 802.11b.

WIM (WAP Identity Module): the security module implemented in a SIM card. The security module is needed for some WAP services, such as banking services or shopping on a WAP site.

Windows CE: a streamlined version of Windows from Microsoft for handheld computers which has since been upgraded and renamed Pocket PC. Windows CE ran Pocket versions of Microsoft office applications such as Word and Excel as well as many applications that were geared specifically for the smaller platform.

wireless: a term used to describe the use of radio-frequency spectrum for transmitting and receiving voice, data and video signals for communications.

Wireless Application Protocol: see WAP.

Wireless Application Service Provider: see WASP.

Wireless Bitmap: see WBMP.

Wireless Communications Services: see WCS.

wireless Internet: a technology that enables a cell phone or other wireless device to access specially formatted Internet content via wireless networks. Several different standards exists: HDML, WML, cHTML, and xHTML. Also known as "Wireless Web" or "WAP".

wireless IP: a packet data protocol standard for sending wireless data over the Internet.

Wireless Local Loop(WLL): wireless service systems that compete with or substitute for local wireline phone service.

Wireless Markup Language: see WML.

wireless modem: see Modem.

wireless operator: a general term that refers to either a wireless network operator, wireless service operator, or a carrier. The wireless network operator maintains the radio towers and infrastructure for a cellular system and sells wireless service to subscribers.

Wireless Personal Area Network: see WPAN.

wireless portal: a web site that supports a user with a smart phone or an alphanumeric pager. It may offer a variety of features, including providing a springboard to other wireless web sites, the ability to select content to be pushed to the user's device as well as providing a point of entry for anyone to send the user a message.

wireless terminal: any mobile phone, wireless handheld, or wireless personal device using non-fixed access to a network.

WISP (Wireless Internet Service Provider): vendors that specialize in providing wireless Internet access to subscribers.

WLAN (Wireless Local Area Network): a network that transmits and receives data over the air using radio frequency technology, minimizing the need for wired connections. A wireless LAN can serve as a replacement for or extension to a wired LAN.

WLIF (Wireless LAN Interoperability Forum): a membership group that endorses products that are interoperable with major standards; supports OpenAir and 802.11.

WLL (Wireless Local Loop): a system that connects subscribers to the public switched telephone network (PSTN) using wireless technology coupled with line interfaces and other circuitry to complete the "last mile" between the customer premise and the exchange equipment. Wireless systems can often be installed in far less time and at lower cost than traditional wired systems.

WML (Wireless Markup Language): a name given to the markup language for WAP. WML is based on XML (HTML's more flexible cousin) and enables information to be displayed on a micro-browser. As a WAP phone cannot process an HTML web page, WML was developed to work within the constraints of narrowband devices.

WMLS (Wireless Markup Language Script): a subset of JavaScript, used to program mobile devices.

world phone: phones that operate on 900, 1800 and 1900 MHZ GSM networks. Because of this, world phones are able to operate in most parts of the world.

WPAN (Wireless Personal Area Network): a wireless network that serves an individual user.

WWW (World Wide Web): one of the primary applications in the Internet. It is a system in which information display is made through the use of hypertext (HTML), where it is possible to combine all Internet services and use text, images and sound simultaneously.

Chapter 51

Glossary of Wireless Telephone Fraud Terms

access fraud: Any unauthorized use of wireless service through the intentional or unintentional tampering, manipulation or programming of a wireless phone's Electronic Serial Number (ESN) and/or Mobile Identification Number (MIN). By replacing or programming stolen, valid ESNs into the chip, thieves can trick the wireless system's computers into thinking that the phone is being used by a legitimate customer.

call selling operations: One of the most costly types of wireless fraud activity is the call selling operation. These unlicensed reseller services are set up to provide international fraudulent wireless communications. By combining the fraudulent phone with a stolen long-distance calling card number, the call seller can offer a low-cost call to people who want to contact their family, friends, or business associates anywhere in the world.

counterfeit "clone" phones: As tumbling fraud diminished, an even more type of fraud appeared. With "ESN Cloning," the criminal doesn't pretend to be an out-of-town roamer but rather a legitimate local customer. With this type of fraud, the criminal puts into a phone a computer

---

"Wireless Telephone Fraud Glossary," Cellular Telecommunications and Internet Association, http://www.wow-com.com/industry/tech/security/articles.cfm?ID=311.

chip that can be programmed with both the ESN and MIN of a legitimate user. The criminal obtains valid number combinations, either through the use of illegally used test equipment or through an unscrupulous employee of a retail agent or carrier.

counterfeit "lifetime" phone: This type of fraud enables thieves to reprogram a special wireless phone through its own keypad so that wireless bills are charged to someone else. With the "lifetime" counterfeit phone technology, numerous legitimate MIN/ESN pairs can be stored in each phone and the electronic component in a wireless phone that matches the ESN with the MIN can be reprogrammed in less than one minute.

counterfeit "tumbler" phones: In late 1990, a new sophisticated type of crime emerged known as "tumbling ESN" fraud. This type of fraud would alter a wireless phone so it would tumble through a series of ESNs and make the caller appear to be another new customer each time a call was made. By replacing the "Number Assignment Module" (NAM), an electronic component in a wireless phone that matches the ESN with the MIN, a typical tumbler caller had the ability to generate 48,000 possible ESNs.

ESN: Electronic Serial Number is a unique number assigned to a wireless phone by the manufacturer. According to the Federal Communications Commission, however, the ESN is to be fixed and unchangeable—a sort of unique fingerprint for each phone.

MIN: Mobile Identification Number assigned by the wireless carrier to a customer's phone. The MIN is meant to be changeable, since the phone could change hands or a customer could move to another city.

NAM: The "Number Assignment Module" (NAM), an electronic component in a wireless phone that matches the ESN with the MIN.

stolen phone fraud: The unauthorized use of a phone stolen from a legitimate customer before that customer can report the theft.

subscription fraud: This type of fraud occurs when a subscriber signs up for service with fraudulently-obtained customer information or false identification without any intention of paying for service.

two phones, one number: Illegal operators which offer consumers the capability to have two wireless phones with one mobile identification number (MIN). These operators offer this service by altering the internal workings of a second wireless phone.

Chapter 52

Public and Private Communications Security and Privacy Organizations

This chapter lists contact information for some of the public and private organizations that deal with various aspects of communications security and privacy. Check individual websites for details. Information is listed alphabetically according to the name of the organization, within each topic.

Cellular/Wireless Phones

American Association for Retired Persons (AARP)
601 E St. NW
Washington, DC 20049
Toll Free: (800) 424-3410
Website: www.aarp.org

See AARP documents "Shopping for Cell Phones" (http://www.aarp.org/confacts/money/cellphone.html), and "Understanding Consumer Use of Wireless Telephone Service" (http://research.aarp.org/consume/d17328_wireless.html).

The resources listed in this section were compiled from a wide variety of sources deemed accurate. Contact information was updated and verified in December 2002. Inclusion does not constitute endorsement.

Wireless Consumers Alliance
P.O. Box 2090
Del Mar, CA 92010
Phone: (858) 509-2938
Fax: (858) 509-2937
Website: www.wirelessconsumers.org
E-mail: mail@wirelessconsumers.org

Nonprofit consumer advocacy organization. Website provides a "what to know" section, as well as information on health and safety concerns, consumer rights, and legal and regulatory actions.

ECHELON

EchelonWatch
Website: www.echelonwatch.org

Electronic Privacy Information Center
1718 Connecticut Avenue NW, Suite 200
Washington, DC 20009
Phone: (202) 483-1140
Fax: (202) 483-1248
Website: www.epic.org
E-mail: info@epic.org

Infosyssec.org
Website: www.infosyssec.org

Information security portal for system security professionals.

Fraud Assistance Organizations

Better Business Bureau
4200 Wilson Blvd., Suite 800
Arlington, VA 22203-1838
Phone: (703) 276-0100
Fax: (703) 525-8277
Website: www.bbb.org

Call for Action
5272 River Road, Suite 300
Bethesda, MD 20816
Phone: (301) 657-8260
Website: www.callforaction.org

National Consumers League (NCL)

1701 K Street NW
Suite 1200
Washington, DC 20006
Phone: (202) 835-3323
Fax: (202) 835-0747
Website: www.nclnet.org

You can order the NCL document "A Survival Guide to Slamming, Cramming, and other Phone Concerns" at http://www.nclnet.org/orderfrm.htm.

National Fraud Information Center

Toll Free: (800) 876-7060
Website: www.fraud.org

Identity Theft

Identity Theft Resource Center

P.O. Box 26833
San Diego, CA 92196
Phone: (858) 693-7935
Website: www.idtheftcenter.org
E-mail: voices123@att.net

The Privacy Rights Clearinghouse

Website: www.privacyrights.org

"Identity Theft: What to Do if It Happens to You" http://www.privacyrights.org/identity.htm.

Long Distance Fraud and Scams

International Prepaid Communications Association

904 Massachusetts Ave. NE
Washington, DC 20002
Toll Free: (800) 958-7824
Phone: (202) 544-4448
Fax: (202) 547-7417
Website: www.i-pca.org

SmartLongDistancePrices.com

Website: www.smartlongdistanceprices.com

SmartPrice.com
8716 N MoPac Expressway
Suite 300
Austin, TX 78759
Website: http://partners.smartprice.com

Maintains an unbiased, current database of information about nearly every long-distance carrier.

Telemarketing Fraud

American Association of Retired Persons (AARP)
601 E St. NW
Washington, DC 20049
Toll Free: (800) 424-3410
Website: www.aarp.org

Better Business Bureau
4200 Wilson Blvd., Suite 800
Arlington, VA 22203-1838
Phone: (703) 276-0100
Fax: (703) 525-8277
Website: www.bbb.org

Direct Marketing Association (DMA)
1120 Avenue of the Americas
New York, NY 10036-6700
Phone: (212) 768-7277 [New York]
Phone: (202) 955-5030 [Washington, DC]
Fax: (212) 302-6714
Website: www.the-dma.org

Internet Fraud Watch
Toll Free: (800) 876-7060
Website: www.fraud.org/internet/intset.htm

National Consumers League
1701 K Street NW, Suite 1200
Washington, DC 20006
Phone: (202) 835-3323
Fax: (202) 835-0747
Website: www.natlconsumersleague.org

National Fraud Information Center (NFIC)

Toll Free: (800) 876-7060

Consumers can contact NFIC to ask about telemarketing calls they receive.

North American Securities Administrators Association

10 G Street NE, Suite 710
Washington, DC 20002
Phone: (202) 737-0900
Fax: (202) 783-3571
Website: www.nasaa.org
E-mail: info@nasaa.org

Total Information Awareness (TIA)

Electronic Privacy Information Center

1718 Connecticut Avenue, NW
Suite 200
Washington, DC 20009
Phone: (202) 637-9800
Fax: (202) 637-0968
Website: www.epic.org
E-mail: info@epic.org

Markle Foundation

10 Rockefeller Plaza
16th Floor
New York, NY 10020-1903
Phone: (212) 489-6655
Fax: (212) 765-9690
Website: www.markle.org

Wireless Data Networks

The Home PC Firewall Guide

Website: www.firewallguide.com/index.htm

Provides access to independent, third-party reviews of Internet security products, including wireless computer networks and PDAs.

PracticallyNetworked.com

Website: www.practicallynetworked.com/support/wireless_secure.htm

Wireless Industry

Cellular Telecommunications and Internet Association
1250 Connecticut Avenue, NW, Suite 800
Washington, DC 20036
Phone: (202) 785-0081
Website: www.wow-com.com
"How Wireless Technology Works"
Webpage: http://www.wow-com.com/consumer/howitworks

The industry association CTIA provides information about the wireless industry, including the latest usage statistics.

Wireless Location Tracking

Center for Democracy and Technology
1634 Eye Street, NW, Suite 1100
Washington, DC 20006
Phone: (202) 637-9800
Fax: (202) 637-0968
Website: www.cdt.org
E-mail: feedback@cdt.org

Electronic Privacy Information Center
1718 Connecticut Avenue, NW
Suite 200
Washington, DC 20009
Phone: (202) 483-1140
Fax: (202) 483-1248
Website: www.epic.org
E-mail: info@epic.org

Location Interoperability Forum
2570 W El Camino Real, Suite 304
Mountain View, CA 94040-1313
Phone: (650) 949-6760
Fax: (650) 949-6765
Website: www.locationforum.org
E-mail: info@mail.openmobilealliance.org

The website listed directs you to the Open Mobile Alliance webpage (www.openmobilealliance.org); the information provided belongs to OMA.

Wireless Location Industry Association
1225 19th Street, NW
Washington, DC 20036-2453
Phone: (202) 955-6067
Website: www.wliaonline.com

Wiretapping

Digital Century
Website: www.digitalcentury.com

"Electronic Communications Privacy Act: Overview," www.digitalcentury.com/encyclo/update/ecpa.html

The National Conference of State Legislatures
444 North Capitol Street, NW
Suite 515
Washington, DC 20001
Phone: (202) 624-5400
Fax: (202) 737-1069
Website: www.ncsl.org/programs/lis/CIP/surveillance.htm

Provides a chart of federal and state electronic surveillance laws.

Ostgate.com

Provides information on bug sweeps, spy hunting, and counterintelligence. See documents "Types of Wiretaps, Bugs and Methods," www.ostgate.com/typebug.html and "Warning Signs of Covert Eavesdropping" www.ostgate.com/warningsigns.html.

Privacy Rights Clearinghouse
Website: www.privacyrights.org

"Wiretapping/Eavesdropping on Telephone Conversations: Is There Cause for Concern?" www.privacyrights.org/fs/fs9-wrtp.htm.

Reporters Committee for Freedom of the Press
1815 N Fort Meyer Drive
Suite 900
Arlington, VA 22209
Phone: (703) 807-2100
Website: www.rcfp.org/taping

Directory of the wiretapping and eavesdropping laws in the 50 states.

Wired.com

"Wiretapping Unwarranted?" http://www.wired.com/news/politics/0,1283,33810,00.html.

Chapter 53

Government Agency Communications Security Resources

This chapter lists contact information for some of the government agencies that deal with various aspects of communications security and privacy. Check individual websites for details. Information is listed alphabetically according to the name of the agency.

CALEA Implementation Section
14800 Conference Center Dr., Suite 300
Chantilly, VA 20151-0450
Toll Free: (800) 551-0336
Phone: (703) 814-4700, Fax: (703) 814-4750
Website: www.askcalea.net

Code of Federal Regulations
Website: www.access.gpo.gov/nara/cfr

Commodity Futures Trading Commission
Three Lafayette Centre
1155 21st Street NW
Washington, DC 20581
Phone: (202) 418-5000, Fax: (202) 418-5521
Website: www.cftc.gov/cftc/
This agency handles telemarketing fraud cases.

The resources listed in this section were compiled from a wide variety of sources deemed accurate. Contact information was updated and verified in December 2002. Inclusion does not constitute endorsement.

The Consumer Action Handbook
Toll Free: (888) 878-3256
Website: www.pueblo.gsa.gov/crh/utility.htm

Your state's public utilities commission also oversees wireless providers and enables you to submit complaints. To find the contact information for your state's utilities commission, consult The Consumer Action Handbook, available free by phone or on the Internet.

Department of Justice
950 Pennsylvania Ave NW
Washington, DC 20530-0001
Phone: (202) 353-1555
Website: www.usdoj.gov
E-mail: AskDOJ@usdoj.gov

This agency handles telemarketing fraud cases.

Federal Bureau of Investigation (FBI)
935 Pennsylvania Ave NW, Room 7972
Washington, DC 20535
Phone: (202) 324-3000
Website: www.fbi.gov/homepage.htm

This agency handles telemarketing fraud cases.

Federal Communications Commission (FCC)
445 12th Street SW
Washington, DC 20554
Toll Free: (888) CALL-FCC [225-5322]
TDD: (888) TELL-FCC [835-5322]
Fax: (202) 418-0232
Website: www.fcc.gov
E-mail: fccinfo@fcc.gov

Read the FCC's guide, "What You Should Know about Wireless Phone Services" www.fcc.gov/cgb/wirelessphone.pdf. Another useful FCC guide is "Market Sense: Cell Phones Fact, Fiction, Frequency," www.fcc.gov/cgb/cell_phones.html. Includes a chart comparing analog and digital wireless phones.

Federal Trade Commission (FTC, including FTC Sentinel)
600 Pennsylvania Ave NW
Washington, DC 20580

Toll Free:(877) IDTHEFT [438-4338], for identity theft; (877) 382-4357, for consumer fraud
Phone: (202) 326-2222
TDD: (202) 326-3128
Websites: www.ftc.gov; www.consumer.gov/idtheft; www.consumer.gov/sentinel
E-mail: sentinel@ftc.gov

The FTC's identity theft clearinghouse. This agency also handles telemarketing fraud cases.

National Association of Attorneys General
750 First Street NE
Washington, DC 20002
Phone: (202) 326-6000
Fax: (202) 408-7014
Webpage: www.naag.org

This agency handles telemarketing fraud cases.

U.S. Customs Service
Customs Headquarters
1300 Pennsylvania Ave NW
Washington DC 20229
Webpage: www.customs.ustreas.gov

This agency handles telemarketing fraud cases.

U.S. Postal Inspection Service
222 Riverside Plaza
Suite 1250
Chicago, IL 60606-6100
Webpage: www.usps.gov/websites/depart/inspect

This agency handles telemarketing fraud cases.

U.S. Securities and Exchange Commission
SEC Headquarters
450 Fifth Street NW
Washington, DC 20549
Phone: (202) 942-8088
TDD: (202) 942-7114
Webpage: www.sec.gov

This agency handles telemarketing fraud cases.

U.S. Sentencing Commission
One Columbus Circle NE
Washington, DC 20002-8002
Phone: (202) 502-4500
Webpage: www.ussc.gov

This agency handles telemarketing fraud cases.

Chapter 54

Privacy Options Resources

This chapter lists contact information for some of the public and private organizations that deal with various aspects of communications privacy. Check individual websites for details. Information is listed alphabetically according to the name of the organization.

BBB OnLine—Understanding Privacy
Website: www.bbbonline.org/understandingprivacy
E-Mail: feedback@cbbb.bbb.org

Offers tips for the consumer and tools and resources to help businesses become more knowledgeable about good privacy practices.

Electronic Privacy Information Center
1718 Connecticut Avenue, NW
Suite 200
Washington, DC 20009
Phone: (202) 483-1140
Fax: (202) 483-1248
Website: www.epic.org/privacy/
E-mail: info@epic.org

Provides news and legislation on First Amendment and constitutional issues of privacy.

The resources listed in this section were compiled from a wide variety of sources deemed accurate. Contact information was updated and verified in December 2002. Inclusion does not constitute endorsement.

Privacy Council
1300 East Arapaho
Suite 300
Richardson, TX 75081
Toll Free: (886) P-Council
Phone: (972) 997-4001
Fax: (972) 997-4450
Website: www.privacycouncil.com

Free Internet library on global privacy law.

Privacy Law Playbook
Website: www.privacylawplaybook.com

Covers privacy law developments that affect business.

Chapter 55

How to File a Complaint with the FCC

The FCC accepts complaints on various wireless and wireline telecommunications issues, media, and telecommunications accessibility issues. The following provides information on how to complain to the FCC.

What Good Will It Do?

Filing an informal complaint with the FCC may help resolve disputes consumers have with companies regulated by the FCC. It will not necessarily result in a fine or enforcement action against the company, but is a way for you, the consumer, to obtain a specific response from the company and, in most cases, a satisfactory resolution to your complaint. After receiving your complaint, FCC staff generally will forward it to the service provider and direct the company to respond to the FCC within 30 days. The FCC also directs the company to send a copy of its response to you (the complainant). If your complaint involves an interstate telephone matter and you do not like the company's response to your complaint, the FCC's rules give you the right to file a "formal" complaint. Consumers who wish to file formal complaints pay a $165.00 filing fee per complaint and must satisfy very specific procedural and evidentiary requirements. For these reasons

---

"Got a Gripe? Filing a Complaint with the FCC Is Easy," Federal Communications Commission (FCC), http://www.fcc.gov/cgb/consumerfacts/complaintfile.html, December 16, 2002.

formal complaints are usually filed by lawyers. For complete information on how to file formal complaints, see 47 CFR Section 1.720 through 1.735. Instructions are also available online at: wireless.fcc.gov/rules.html.

Types of Complaints the FCC Handles

The FCC handles a variety of complaints, including but not limited to:

- state-to-state (interstate) long distance telephone service;
- cellular service;
- paging;
- telephone and equipment accessibility (for persons with disabilities);
- unwanted telemarketing calls;
- obscene and indecent material broadcast over the airwaves;
- technical matters like frequency, antenna registration, interference and tower lighting;
- closed captioning and access by hearing impaired to emergency information on television in your home; and
- hearing aid compatibility of telephones, including payphones and wireless devices.

The FCC does not regulate information services (computers or the Internet). It also does not handle complaints relating to these types of services.

It's Convenient—and It's Free!

The FCC accepts complaints in a number of ways:

Phone. Call us with your complaint. Operators are available M–F, 8 am–5:30 pm EST.

1-888-CALL-FCC (1-888-225-5322) voice

1-888-TELL-FCC (1-888-835-5322) TTY

E-mail. E-mail your complaint to: fccinfo@fcc.gov

Mail. Send your complaint to:

Federal Communications Commission (FCC)
Consumer & Governmental Affairs Bureau
Consumer Complaints
445 12th Street, SW
Washington, DC 20554

Fax. Fax your complaint to: (202) 418-0232

Electronically. For complaints about wireless or wireline phone-related issues, file online through our website at: www.fcc.gov/cgb/complaints.html.

What to Include

Your complaint should include the following information:

- your name, address and the telephone number or numbers involved with your complaint (if telephone-related);
- a telephone number where you can be reached during the business day;
- specific information about your complaint, including the names of all companies involved;
- names and telephone numbers of any company representatives you contacted, dates you spoke with these representatives and any other information that would help process your complaint;
- a copy of any bills which relate to the dispute; and
- the type of resolution you are seeking, such as a credit or refund.

Index

Index

Page numbers followed by 'n' indicate a footnote. Page numbers in *italics* indicate a table or illustration.

A

D

F

G

H

I

J

K

L

M

P

R

S

U

V

W

Z